Meeting the Energy Needs of FUTURE WARRIORS

Committee on Soldier Power/Energy Systems
Board on Army Science and Technology
Division on Engineering and Physical Sciences

NATIONAL RESEARCH COUNCIL
OF THE NATIONAL ACADEMIES

THE NATIONAL ACADEMIES PRESS
Washington, D.C.
www.nap.edu

THE NATIONAL ACADEMIES PRESS • 500 Fifth Street, N.W. • Washington, DC 20001

NOTICE: The project that is the subject of this report was approved by the Governing Board of the National Research Council, whose members are drawn from the councils of the National Academy of Sciences, the National Academy of Engineering, and the Institute of Medicine. The members of the committee responsible for the report were chosen for their special competences and with regard for appropriate balance.

This study was supported by Contract/Grant No. DAAD19-03-C-0046, between the National Academy of Sciences and the Department of the Army. Any opinions, findings, conclusions, or recommendations expressed in this publication are those of the author(s) and do not necessarily reflect the views of the organizations that provided support for the project.

International Standard Book Number 0-309-09261-2 (Book)
International Standard Book Number 0-309-53344-9 (PDF)

Additional copies of this report are available from the National Academies Press, 500 Fifth Street, N.W., Lockbox 285, Washington, DC 20055; (800) 624-6242 or (202) 334-3313 (in the Washington metropolitan area); Internet, http://www.nap.edu

Printed in the United States of America

THE NATIONAL ACADEMIES

Advisers to the Nation on Science, Engineering, and Medicine

The **National Academy of Sciences** is a private, nonprofit, self-perpetuating society of distinguished scholars engaged in scientific and engineering research, dedicated to the furtherance of science and technology and to their use for the general welfare. Upon the authority of the charter granted to it by the Congress in 1863, the Academy has a mandate that requires it to advise the federal government on scientific and technical matters. Dr. Bruce M. Alberts is president of the National Academy of Sciences.

The **National Academy of Engineering** was established in 1964, under the charter of the National Academy of Sciences, as a parallel organization of outstanding engineers. It is autonomous in its administration and in the selection of its members, sharing with the National Academy of Sciences the responsibility for advising the federal government. The National Academy of Engineering also sponsors engineering programs aimed at meeting national needs, encourages education and research, and recognizes the superior achievements of engineers. Dr. Wm. A. Wulf is president of the National Academy of Engineering.

The **Institute of Medicine** was established in 1970 by the National Academy of Sciences to secure the services of eminent members of appropriate professions in the examination of policy matters pertaining to the health of the public. The Institute acts under the responsibility given to the National Academy of Sciences by its congressional charter to be an adviser to the federal government and, upon its own initiative, to identify issues of medical care, research, and education. Dr. Harvey V. Fineberg is president of the Institute of Medicine.

The **National Research Council** was organized by the National Academy of Sciences in 1916 to associate the broad community of science and technology with the Academy's purposes of furthering knowledge and advising the federal government. Functioning in accordance with general policies determined by the Academy, the Council has become the principal operating agency of both the National Academy of Sciences and the National Academy of Engineering in providing services to the government, the public, and the scientific and engineering communities. The Council is administered jointly by both Academies and the Institute of Medicine. Dr. Bruce M. Alberts and Dr. Wm. A. Wulf are chair and vice chair, respectively, of the National Research Council.

www.national-academies.org

COMMITTEE ON SOLDIER POWER/ENERGY SYSTEMS

PATRICK F. FLYNN, NAE, *Chair*, Cummins Engine Company, Inc. (retired), Columbus, Indiana
MILLARD F. ROSE, *Vice Chair*, Radiance Technologies, Huntsville, Alabama
ROBERT W. BRODERSEN, NAE, University of California at Berkeley
ELTON J. CAIRNS, Lawrence Berkeley National Laboratory, Berkeley, California
HUK YUK CHEH, Duracell, Bethel, Connecticut
WALTER L. DAVIS, Motorola Corporation, Schaumburg, Illinois
ROBERT H. DENNARD, NAE, Thomas J. Watson Research Center, IBM, Yorktown Heights, New York
PAUL E. FUNK, U.S. Army (retired), University of Texas at Austin
ROBERT J. NOWAK, Defense Advanced Research Projects Agency (retired), Silver Spring, Maryland
JEFFREY A. SCHMIDT, Ball Aerospace & Technologies Corporation, Boulder, Colorado
DANIEL P. SIEWIOREK, NAE, Carnegie Mellon University, Pittsburgh
KAREN SWIDER LYONS, Naval Research Laboratory, Washington, D.C.
ENOCH WANG, Central Intelligence Agency, McLean, Virginia
DONALD P. WHALEN, U.S. Army (retired), Cypress International, Arlington, Virginia

Board on Army Science and Technology Committee Advisor

ALAN H. EPSTEIN, NAE, Massachusetts Institute of Technology, Cambridge

Staff

BRUCE A. BRAUN, Director, Board on Army Science and Technology
ROBERT J. LOVE, Study Director
DANIEL E.J. TALMAGE, JR., Research Associate
TOMEKA N. GILBERT, Senior Program Assistant

BOARD ON ARMY SCIENCE AND TECHNOLOGY

Staff

Preface

The Army's future force will continue to be based on highly capable dismounted soldiers. The success of these future warriors will depend on enhanced situational awareness, that is, detailed knowledge of the location and capabilities of both friendly and enemy forces, and on improved access to lethal weapons, including those that might be called upon from supporting forces. To enable the transition to such a future force, the soldiers' uniforms, weapons systems, sensors, and communication capabilities are all going through a period of revolutionary development. Perhaps the most critical of these new developments are power supply systems to allow the new electronics-based equipment to function effectively for missions up to 72 hours in length.

Ensuring adequate power for soldiers on the battlefield is by no means a simple problem; otherwise, the Army would not have asked the National Research Council (NRC) to do this study. It is a multidimensional challenge requiring multidimensional approaches, and the solutions involve a full consideration of power/energy systems, including the energy sources, energy sinks, and energy management.

Developers of the original Land Warrior suite of equipment grappled with shortcomings in power as well as the relative immaturity of computer and electronics technologies. Future soldiers, operating in concert as part of a light and mobile force, will depend heavily on networked applications for both situational awareness and access to supporting fires. As a consequence, power for communications-electronics will become the most critical component of warrior capabilities.

Each new capability brings with it a claim on existing weight and space to be borne by the dismounted soldier. For the soldier to function effectively, these weight and space assertions must be limited. Key to this management process will be controlling power demand and providing the power and energy systems that place minimal weight and space demands on the soldier.

With a vision of the Future Force warrior provided by the Army, as well as the results of previous studies on the subject, the NRC Committee on Soldier Power/Energy Systems was chartered by the Army to review the state of the art and recommend technologies that will support the rapid development of effective power source systems for soldier applications. The committee was also asked to review opportunities and technologies for reducing and managing power use. To accomplish this, the committee members necessarily represented a broad range of technical expertise, from computers, communications, low-power electronics, and multiple areas of energy sources, to military logistics, operations, and training. (See Appendix A for biographies of the committee members.)

I would like to express my personal appreciation to the committee members for their helpful and objective participation in reviewing the status of technologies and programs and in recommending directions for future activities. This report is the product of their efforts and consensus. I would also like to express the committee's appreciation to the NRC staff for the large logistic and administrative effort that was required to complete the report.

Patrick F. Flynn
Chair, Committee on
Soldier Power/Energy Systems

Acknowledgment of Reviewers

This report has been reviewed in draft form by individuals chosen for their diverse perspectives and technical expertise, in accordance with procedures approved by the Report Review Committee of the National Research Council (NRC). The purpose of this independent review is to provide candid and critical comments that will assist the institution in making its published report as sound as possible and to ensure that the report meets institutional standards for objectivity, evidence, and responsiveness to the study charge. The review comments and draft manuscript remain confidential to protect the integrity of the deliberative process. We wish to thank the following individuals for their review of this report:

Henry W. Brandhorst, Auburn University,
Douglas M. Chapin, MPR Associates, Inc.,
Bruce S. Dunn, University of California at Los Angeles,
David E. Foster, University of Wisconsin,
Samuel Fuller, Analog Devices,
Gilbert Herrera, Sandia Laboratories,
Nguyen Minh, General Electric Hybrid Power Generation Systems,
Leon E. Salomon, U.S. Army (retired),
Clarence G. Thornton, Army Research Laboratory (retired), and
Robert Whalin, Jackson State University.

Although the reviewers listed above have provided many constructive comments and suggestions, they were not asked to endorse the conclusions or recommendations nor did they see the final draft of the report before its release. The review of this report was overseen by Alton D. Romig, Jr., Sandia National Laboratories, who was responsible for making certain that an independent examination of this report was carried out in accordance with institutional procedures and that all review comments were carefully considered. Responsibility for the final content of this report rests entirely with the authoring committee and the institution.

Contents

APPENDIXES

Figures and Tables

FIGURES

TABLES

Acronyms and Abbreviations

ACRONYMS

AMTEC	alkali metal thermal to electrical conversion
ARL	Army Research Laboratory
ASB	Army Science Board
ASIC	application-specific integrated circuit
ATD	advanced technology demonstration
BOP	balance-of-plant
CECOM	Communications-Electronics Command
CMOS	complementary metal-oxide semiconductor
CO	carbon monoxide
COTS	commercial off-the-shelf
CPOX	catalytic partial oxidation
CPU	central processing unit
DARPA	Defense Advanced Research Projects Agency
DMFC	direct methanol fuel cell
DOD	U.S. Department of Defense
DOE	U.S. Department of Energy
DRAM	dynamic random access memory
DSP	digital signal processing
EC	electrochemical capacitor
EOD	end of discharge
FPGA	field programmable gate array
GPS	Global Positioning System
HHV	higher heating value
HIA	high integration actuator
HMMWV	high-mobility multipurpose wheeled vehicle
HPC	high performance computing
HUD	heads-up displays

IC	internal combustion
IEEE	Institute of Electrical and Electronics Engineers
IP	Internet Protocol
JP	jet propellant
JTRS	Joint Tactical Radio System
LHV	lower heating value
LLNL	Lawrence Livermore National Laboratory
LTI	lead technology integrator
LW	Land Warrior
LW-AC	Land Warrior-Advanced Capability
LW-SI	Land Warrior-Stryker Interoperable
MBITR	multiband intra/inter team radio
MCC	microclimate cooling
MEA	membrane electrode assembly
MEMS	microelectromechanical systems
MIMO	multiple-input, multiple-output
MURI	Multidisciplinary University Research Initiative
NASA	National Aeronautics and Space Administration
NRC	National Research Council
NTRS	National Technology Roadmap for Semiconductors
OCV	open circuit voltage
OFW	Objective Force Warrior (aka Future Force Warrior)
PAN	primary area network
PC	personal computer
PEM	proton exchange membrane
PEMFC	proton exchange membrane fuel cell
PEO	Program Executive Office
PMMEP	Project Manager Mobile Electric Power
R&D	research and development
RF	radio frequency
S&T	science and technology
SI	Stryker Interoperable
SIA	Semiconductor Industry Association
SOA	state of the art
SoC	system-on-a-chip
SOF	special operations forces
SOFC	solid oxide fuel cell
SRAM	static random access memory
TE	thermoelectrics
TPV	thermophotovoltaics
TRADOC	Training and Doctrine Command
TRL	technology readiness level
UAW	universal access workstation
UWB	ultrawideband

VGA	video graphics array
VTB	virtual testbed
WLAN	wireless local area network
YSZ	yttria-stabilized zirconia

ABBREVIATIONS

μm	micrometer
A	ampere
Ah	ampere-hour
Al/air	aluminum/air
C	coulomb
C/air	carbon/air
cc	cubic centimeter
Cd/NiOOH	cadmium/nickel
$(CF)_x$	carbon monofluoride
dB	decibel
g	gram
GHz	gigahertz
hp	horsepower
I	current
J	joule
kg	kilogram
kJ	kilojoule
kW	kilowatt
kWh	kilowatt-hour
L	liter
Li	lithium
Li/air	lithium/air
$Li/(CF)_x$	lithium/carbon monofluoride cell
$LiCoO_2$	lithium cobalt oxide
$LiFePO_4$	lithium iron phosphate
Li/MnO_2	lithium/manganese dioxide
$LiMn_2O_4$	lithium manganese oxide
$LiNiO_2$	lithium nickel oxide
Li/S	lithium/sulfur
Li/SO_2	lithium/sulfur dioxide
MeOH	methanol
Mg	magnesium
MH/NiOOH	nickel/metal hydride

MHz	megahertz
MIPS	million instructions per second
mJ	millijoule
MKS	meter-kilogram-second
mol	mole
MOPS	million operations per second
mW	milliwatt
$NaBH_4$	sodium borohydride
nm	nanometer
ppm	parts per million
psi	pounds per square inch
PvdF	polyvinylidene fluoride
V	volt
W	watt
W/cc	watts per cubic centimeter
W/g	watts per gram
Wh	watt-hour
Wh/cc	watt hours per cubic centimeter
W/kg	watts per kilogram
W/L	watts per liter
Zn/air	zinc/air

Executive Summary

Soldier requirements for power are changing as fast as new electronics are being developed. In addition to communications and computers, a myriad of applications for the dismounted soldier of the future will require portable energy, including such things as laser-designators, chemical-biological sensors, uniform ventilators, and exoskeletal enhancements. This report assesses power/energy sources, low-power electronics, and power management technologies and provides recommendations on energy solutions for the future soldier. The committee focused on realistic energy alternatives, concentrating on the energy source technologies about which enough data were available to support likely system concepts and to estimate essential system parameters. The report builds on technology assessments documented in Appendix D and in a previous NRC report, *Energy-Efficient Technologies for the Dismounted Soldier* (NRC, 1997), which will be referred to throughout this report as *Energy-Efficient Technologies.*

Electronics are critical to soldier combat effectiveness. Primary batteries now provide the main energy source, but the acquisition, storage, distribution, and disposal of over a hundred different battery types poses an enormous logistical challenge on the battlefield. New technologies have at the same time increased the number and variety of power-driven functions that require soldier-portable power. In the early 1980s, the Army recognized that it must approach equipping the dismounted soldier from an integrated system vantage. The concept of the soldier as a system led to a prototype of the first Land Warrior (LW) system, which combined electronics, weapons, and power sources in a single ensemble.

BACKGROUND

Dismounted soldiers act as both sensors and shooters, and the LW suite of electronics enhances combat effectiveness through increased situational awareness. Night-vision and infrared sights extend the reach of personal weapons, computer displays provide maps and locations of friendly and enemy troops, communications send and receive information about prospective targets as well as available sources of fire beyond rifle range. Suitably equipped soldiers can relay details about local targets and bring to bear virtually unlimited firepower, a capability that would have been inconceivable as recently as the first Gulf War.

But these capabilities come at a cost. Even without the LW equipment, the physical load borne by a dismounted soldier can exceed 100 pounds for certain missions. When it is fielded, the weight of the LW ensemble may add 30 pounds or more, not counting any extra batteries needed to guarantee power for the mission, which would clearly impact the soldier's combat effectiveness.

The Army Program Executive Officer-Soldier is responsible for both the LW acquisition program and the Objective Force Warrior-Advanced Technology Demonstration (OFW-ATD) program.[1] The OFW-ATD is working to integrate LW electronics using advanced concepts and to demonstrate an OFW prototype in 2004. This may serve as the basis for a future generation of LW to be fielded in the 2007-2010 time frame. In the far term, the Army envisions integrating soldier functions even more extensively by possibly embedding the electronics in a uniform made of advanced materials.

Portable power/energy sources were reviewed and the power demand was categorized in distinct regimes at the Energy and Power Workshop for the Soldier, sponsored by the Army Research Laboratory/Communications Electronics Command (ARL/CECOM). This workshop and the earlier National Research Council study, *Energy-Efficient Technologies* (NRC, 1997), provided the foundations for this study, which was requested by the Assistant Secretary of the

[1]Since completion of the study, the term "Future Force" has been adopted by the Army in place of "Objective Force."

Army (Acquisition, Logistics, and Technology) to accomplish the following tasks:

1. Expand upon the conclusions from the ARL/CECOM Energy and Power Workshop for the Soldier, held on 15-17 October 2002, through the specification of both impact and feasibility of incorporating power management components, techniques and procedures for powering low-power electronic devices. The specific regimes from the workshop were: 20-watt average with 50-watt peak and 100-watt average with a 200-watt peak for up to 72-hr missions. Address power for high-power draw applications such as exoskeleton applications (1 to 5 kW average).
2. Assess electric power technologies to support soldier applications associated with future power and energy demands on the battlefield, e.g. expected OFW operational capabilities for the 2005-2025 time frame, with emphasis on alternative compact high-power and energy-dense sources, power management and distribution techniques, and low-power electronics such as asynchronous microchips, smart dust, etc. Assess technical risks and feasibility associated with each of the technologies and make recommendations pertaining to their potential efficacy and utility within the context of future OFW operational capabilities. Consider risks associated with technology development, integration of hybrid generators and sources, adaptation of commercial technologies, and battlefield logistics. Systems concepts involving appropriate power sources, power management and low-power electronics are to be specified and delineated.
3. Update the technologies evaluated in the 1997 NRC report on Energy-Efficient Technologies for the Dismounted Soldier including changes in individual technology development trends. Determine advantages and disadvantages for appropriate technologies in prospective application areas. Develop standard measures to facilitate comparison.
4. Prepare a consensus report documenting the study results and containing findings and recommendations to assist the Army in its development program. Prioritize the energy source alternatives appropriate to each application. Propose science and technology (S&T) objectives leading to the future incorporation in the Objective Force Warrior program. The report will include:

   a. Recommendations for examined technologies with high benefit for target regimes with detailed justification for technology selection or rejection.
   b. Recommendations for power distribution techniques for soldier systems. Applicability of low-power electronics, such as asynchronous microchips, smart dust, etc., to soldier device loads.
   c. Recommendations for centralized vs. distributed power management for soldier systems including software/hardware techniques for control and conversion.
   d. Applicability of examined technologies to single type sources vs. hybrid sources considering logistics, versatility, utility, environmental factors, safety, reliability, logistic infrastructure, manufacturability and availability.
   e. Recommendation for recharging from soldier carried sources, robots (or vehicle) or fixed platforms.
   f. Recommendations for predictive models and modeling techniques that would elucidate power use and management.

This executive summary summarizes key findings, including the science and technology (S&T) objectives in compact high-power and energy-dense source technologies for each of the regimes, and enumerates the specific recommendations contained in the study report.

TECHNOLOGY ASSESSMENTS

Consistent with the ARL/CECOM Workshop, the committee assumed that the 20-W regime included power solutions for computers, radios, sensors, displays—all electronics subsystems of the LW ensemble. The 100-W regime included niche applications such as high-demand laser target designators and future microclimate cooling capabilities. Finally, the 1- to 5-kW regime was assumed to include the most power-intensive capabilities, such as portable power generators, rechargers for rechargeable batteries, and future exoskeleton devices.

The committee assessed and compared technologies at varying levels of technology readiness. Energy per unit of system mass, i.e., specific energy, served as the primary metric for selecting the technologies with greatest potential for Army purposes from among the many alternatives. Three important issues had to be addressed to make valid comparisons. First, the total energy produced must be measured under identical load conditions (power profile). Second, since fully packaged systems are not available for many of the emerging technologies, comparable parameters had to be estimated. Third, since batteries specify different performance specifications for different cell sizes, the committee provided varying allowances for packaging.

Fueled systems, which are in various stages of development, can be used to replace batteries as well as to supplement batteries in a hybrid system; the committee calculated standard mission energy requirements and used these to compare required masses for battery and fueled systems. Such things as fuel tank and fuel, energy content of the fuel, and energy conversion efficiency were used to compute comparable performance metrics.

Based on these considerations, the committee evaluated

and selected technologies with the greatest potential in each regime. (See Recommendations 6 through 8.) It also developed S&T objectives for the Army consistent with these recommendations, as shown in Table ES-1 for the near term (2010), medium term (2015), and far term (beyond 2015). Table ES-1 also indicates the relative risk (low, medium, or high) associated with each objective. Technologies considered as viable alternatives had to have demonstrated a level of technology readiness that would enable the committee to estimate its performance in a power/energy source system. Because of this, the Army will need to conduct detailed trade studies (specific energy vs. logistics, signature, cost, and so forth) to confirm that particular power source solutions are suited for particular applications.

TABLE ES-1 Science and Technology Objectives for the Near Term, Mid-Term, and Far Term, in Three Power Regimes

Power Regime	Near Term (3 to 5 years)	Mid-term (5 to 10 years)	Far Term (beyond 10 years)
20 W average power	Develop batteries for the 24-hr mission with specific energies >300 Wh/kg.	Develop rapid start-up, compact solid oxide fuel cell (SOFC) systems operating on low-sulfur logistics fuel or surrogates.	Develop high-specific-energy, air-breathing battery system hybrids.
	Develop smart hybrid systems with high-energy and high-power batteries and/or electrochemical capacitors.	Develop complete small internal combustion and Stirling engine systems with low signatures operating on JP-8 or diesel fuels.	Develop microelectromechanical system components for power technologies.
	Develop generic modeling capabilities.		Develop SOFC systems that operate directly on high-sulfur and polyaromatic fuels.
	Develop efficient balance-of-plant components for small fuel cell systems.		
	Develop small fuel processors for logistics fuel, methanol, ammonia, and other viable fuels.		
	Develop and field-test direct methanol fuel cell (DMFC) hybrid systems.		
	Develop and field-test proton exchange membrane/hydrogen (PEM/H_2) systems.		
	Conduct battlefield-relevant safety testing of alternatives (H_2, MeOH, ammonia, JP-8, and Li batteries).		
100 W average power	Develop smart hybrid systems with small engines and fuel cells.	Develop small engines. Validate performance scaling laws. Assess reliability, failure modes.	Develop high-specific-energy, air-breathing batteries.
	Develop portable fuel processors for logistics fuel.	Develop SOFCs.	
	Evaluate DMFC and PEM systems for various specific missions.		
1 to 5 kW average power	Develop lightweight, efficient, 1- to 5-kW engines that operate on logistics fuel.	Integrate logistics fuel reformers with lightweight PEM fuel cells.	Develop high-capacity SOFCs and integrate them with logistics fuel reformers.
	Develop lightweight logistics fuel reformers.		

KEY: Relative risk: Low, ☐ ; Medium, ☐ ; High, ☐ .
NOTES: MeOH, methanol; JP-8, jet propellant 8; Li, lithium.

Battery and Fuel Cell Development

Batteries are the generic solution for soldier power. They will be an integral part of hybrid and stand-alone energy sources for the foreseeable future. The challenge is to make them smaller, lighter, cheaper, more reliable, and more energy-dense without sacrificing safety. Fuel cells are the focus of intense interest by the military because of their potential as instantly "rechargeable" energy sources that can meet specific energy requirements for high electrical loads and long mission lengths. Like metal/air batteries, fuel cells are air-breathing devices that cannot operate when submerged in water. Future acceptance of fuel cells on the battlefield will be determined to a great degree by logistics, because current prototypes are fueled by the nonstandard logistics fuels (methanol and hydrogen).

Recommendation 1: The Army should focus on batteries with a specific energy of 300 Wh/kg and higher for insertion into future versions of the Land Warrior (LW) ensemble. It should continue to promote and support innovative approaches to disposable and rechargeable batteries that can be adapted for military use. To select the best candidates for a given application, the Army should explore the trade-off space that exists between lifetime (measured in terms of charge-discharge cycles), specific power, specific energy, safety, and cost.

Logistical and Operational Considerations

Batteries currently add a substantial burden to the heavy load carried by the dismounted soldier. Use of disposable batteries in training and field operations has proven to be a substantial expense. Employment of rechargeable batteries for many applications promises to reduce life cycle cost but adds the cost of additional equipment and the logistics complexity of recharging in forward areas. Fueled hybrid solutions offer even greater promise than rechargeable batteries in reducing weight for longer missions. These have operational advantages and limitations but add tasks for the logistician, who would have to deal with another nonstandard fuel to be carried forward.

Recommendation 2: The Army should evaluate the applicability of small-scale, portable fuel processors capable of reforming the Army-standard fuels for use in proton exchange membrane (PEM) fuel cells or solid oxide fuel cells (SOFC). Scaling laws should be determined and cost/benefit analyses should be performed to determine whether there are power levels and/or mission durations that make such reformers an attractive alternative.

The Army must determine whether an alternative, nonstandard fuel source (such as methanol, hydrogen, or ammonia) is logistically acceptable. A proper analysis of trade-offs would permit decision makers to make an informed judgment on whether the operational advantages outweigh added logistics complexity and costs. Ideally, this would include testing in line units (even if only at the squad level) under representative field conditions. It would also save the Army money otherwise invested in research on fueled system alternatives that do not make logistical or operational sense.

Recommendation 3: The Army should immediately conduct a comprehensive and definitive analysis of the operational and logistical implications of fielding nonbattery solutions as power sources for dismounted soldiers. This should include consideration of operational benefits, logistical limitations, and life-cycle costs, as well as considerations of safety and risk. It should develop models of competing energy sources, including fuel cell systems, and use them in simulations of battlefield operations. The data can then be combined with estimates of system costs to conduct cost/benefit analyses that would either support the consideration of non-standard-fueled fuel cell systems or eliminate them from consideration.

Small Engines

Several internal and external combustion engine prototypes have been demonstrated and show potential for military applications. Microturbines have not to date demonstrated the ability to provide a net positive system power output. Stirling engines use standard logistics fuel (JP-8) and could serve as a power source for battery rechargers or to meet anticipated requirements for high-demand microclimate cooling and exoskeletal applications. All small internal combustion engine systems now available have distinctive acoustic and heat signatures that would restrict their utility in combat. Stirling engines are inherently quiet but have significant thermal signatures.

Recommendation 4: The Army should adjust the focus of internal combustion engine development to demonstrate net power outputs and balance-of-plant systems appropriate to specific Army applications. Heavy emphasis should be placed on developing packaged systems with reduced heat and noise signatures. Once power output capabilities are demonstrated, the development should focus on improving system efficiencies.

Hybrid Power Systems

Hybrids offer enormous advantages from a simple energetics point of view for longer mission times. A hybrid power/energy system can be optimized for both high energy and high power demands. It can also provide the means to overcome the disadvantage of an air-breathing power source by combining an air-breathing system (e.g., metal/air battery, fuel cell, small engine) with a rechargeable battery.

To be acceptable for soldier use, a power/energy source must be impervious to dust and moisture. An acceptable fueled hybrid must be smart; that is, it must be capable of sensing and reacting to its environments so as to allow the unit to operate under water and to protect it from destruction. Modeling is critical to the design of acceptable hybrid systems.

Recommendation 5: The Army should refine duty-cycle estimates for the Land Warrior suite of electronics so as to enable the development of high-fidelity models incorporating soldier usage patterns and other details of interactions between power sources and soldier electronics. These estimates are essential for developing smart hybrid systems that can react to the environment for the future LW as well as for developing energy-efficient systems to meet unforeseen Army mission requirements.

Technologies for Target Regimes

While many commercial energy sources exist, they are driven by the consumer market and are not developed in sizes commensurate with the broad spectrum of Army needs. The committee was specifically requested in the task statement to select and prioritize power source alternatives in each of the three target regimes. Recommendations 6 through 8 are consistent with the previous recommendations and the S&T objectives set for the Army in Table ES-1.

20-W Average with 50-W Peak

Recommendation 6a: As its first priority in the 20-W target regime, the Army should support development of batteries with specific energies greater than 300 Wh/kg (e.g., Li/$(CF)_x$, Li/S, Li/air, C/air) in sizes commensurate with LW requirements.

Recommendation 6b: The Army should develop smart hybrid systems capable of air-independent operation and the 50-W peak load. These hybrid systems must be developed with the aid of duty-cycle analysis and modeling. Key to this is an evaluation of the limits of battery-battery hybrid system performance as well as methods for packaging or sealing air-breathing hybrid systems.

Recommendation 6c: If the Army determines that a nonstandard fuel source is acceptable for battlefield use by dismounted soldiers (see Recommendation 2 above), it should develop PEM and SOFCs as complete systems with the hydrogen storage or generation subsystem yielding at least 6 percent by weight hydrogen, including all components. In this context the Army should investigate methods of reforming methanol, ammonia, butane, and liquid hydrocarbon fuels and should evaluate whether the development of direct methanol fuel cell (DMFC) systems would be less complex than fuel-processing approaches.

Recommendation 6d: As a final priority in the 20-W regime, and for the far term, the Army should develop and evaluate small engines that operate on standard logistics fuels.

100-W Average with 200-W Peak

Recommendation 7a: As its first priority in the 100-W target regime, the Army should develop smart hybrid systems capable of air-independent operation that can accommodate total energy requirements. The emphasis should be placed on fueled systems (small engines, fuel cells) capable of operating on standard logistics fuels.

Recommendation 7b: The Army should support development of high-specific-energy batteries for niche applications, such as laser designators.

1- to 5-kW Average

Recommendation 8a: As its top priority in the 1- to 5-kW regime, the Army should continue to develop lightweight engines with high specific power that operate on standard logistics fuels. It should investigate Stirling engines, as they are fuel-versatile and offer significant acoustic signature reduction.

Recommendation 8b: For the 1- to 5-kW regime, the Army should develop the ability to process standard logistics fuels as needed for emerging high-specific-power PEM and solid oxide fuel cells.

LAND WARRIOR SYSTEM

Correctly matching power source technologies (sources) with particular electronics applications (sinks) can greatly affect energy efficiency. System developers must also consider how Army logistics and operations impact the selection of power solutions for the soldier. The duty cycle is extremely important when considering a hybrid power solution. Also, dismounted soldiers who are accompanied in combat by a robotic vehicle, as envisioned by the Army for the future, will have a possible means for recharging batteries or fuel supplies that soldiers operating alone do not have.

Considering the OFW prototype as a possible third generation of LW electronics, the average power has been estimated at 20 W and the peak power at 60 W, for all three generations. The committee observed that power savings made possible by technology improvements in later electronics designs, primarily in computer processors, have been traded for improved combat effectiveness as well as to allow the use of plug-and-play architecture to support future evolution. While the desire for such flexibility is understood, the approach comes at a high energy cost and restricts the use of more energy-efficient design solutions.

Energy-Efficient Technologies (NRC, 1997) determined that a LW system averaging only 2 W would be possible if commercial design approaches, including system-on-a-chip (SoC) technology, could be applied to developing the soldier system. Use of SoC design techniques could reduce power by over an order of magnitude for the digital computing and communications processing, making it negligible in comparison with the power demand of analog, sensor, and display functions.

Commercial progress in developing low-power technology has been rapid, even outstripping Semiconductor Industry Association (SIA) roadmap estimates. Since 1997, the energy efficiency of circuits has improved by at least a factor of five. By one measure, this improvement in reducing power demand is greater than the improvement in rechargeable batteries, since time between recharges has increased only 20 percent. There are barriers to continuing improvements, the most important being the power lost due to leakage currents, but the Army has yet to avail itself of any of the gains made in past years.

Reducing power demand is an Army concern, but it is drowned out by the Army's relatively near-term objectives to field and upgrade successive versions of LW. The committee believes that neither the LW acquisition program nor the OFW-ATD programs are large enough or have long enough development horizons to deal effectively with power issues. The simple fact is that laws of physics, chemistry, and size are unlikely to produce the required near-term gains in energy, weight, and size of wearable power sources that will be needed while maintaining the current agility of the soldier. It is therefore imperative that the Army devote R&D effort to reduce the power drain in parallel with continued development and improvement of power sources.

Both the LW acquisition and the OFW-ATD programs rely on other Army programs to develop and acquire the component electronics. None of these programs have an incentive to develop or procure electronics using commercially proven design approaches to reduce energy consumption. And, because of the added cost and risk involved in development, there are actually disincentives for reducing power demand.

As tempting as it may be for the Army to continue use of traditional design techniques, a different strategy is required to design the equipment that the soldier must carry as compared with equipment for vehicles or other mobile or fixed platforms. Consider that there are major differences between what is required to design a smart cell phone and what was required to design an office telephone or home computer. Just as cell phone users have special requirements, the soldier is a unique platform on which must be built a complex electronics system. For these reasons, it is important for the Army to increase its investment in Land Warrior electronics sufficient to begin a customized system-on-a-chip (SoC) approach to the development of future warrior systems. Achieving energy efficiency for these electronics will resolve a myriad of problems now associated with the integration of disparate systems in addition to reducing soldier energy needs.

The Army acquisition system is impaired in its ability to focus on soldier power issues, because it does not take into account the logistics costs of providing power on the battlefield when computing the true life-cycle costs of soldier electronics. The Army should take advantage of the new power-reduction designs and techniques that are well known in commercial industry, especially in light of the stakes involved with future soldiers on the battlefield.

Recommendation 9: The Army should make realistic estimates of the life-cycle cost, including reasonable logistics costs, of providing power on the battlefield and use such estimates in determining how much to invest in future Land Warrior design and development. Additional funding to extend the technology horizon of the program would enable a design solution that optimizes low-energy applications.

Power for Soldier Communications

Wireless communications is the most power-hungry of soldier electronics applications and offers the best chance to reduce future warrior energy requirements. The importance of focusing on communications-electronics was emphasized in *Energy-Efficient Technologies* (NRC, 1997), but the Army has yet to pay attention. Five years later, the power performance planned by OFW-ATD for the Joint Tactical Radio System (JTRS) soldier radio is based on a rough equivalence with the MBITR radio, hardly the cutting edge of energy-efficient radios.

There is clear evidence that reductions in power demand are not a high enough priority for communications-electronics. Power and duty-cycle estimates for the LW soldier radio have not been refined for at least 5 years, even though communications technology has advanced considerably and new network-centric capabilities are planned to one day connect every soldier on the battlefield.

Recommendation 10: The Army should make energy efficiency a first-order design parameter whenever specifying system performance parameters in its contracts. It should provide monetary incentives as needed to reduce power demand in all its procurements for soldier electronics, especially for communications.

OVERARCHING RECOMMENDATIONS

The OFW focus on increasing combat effectiveness rather than energy efficiency encourages trading off power savings achieved for new electronics. With no net reduction in power, this approach could undermine the benefits of a system-of-systems design approach to reducing power demand and could contribute further to the chasm that exists between the consumer electronics' state of the art and Army state of the art.

Table ES-2 summarizes areas within the Land Warrior system that are key to improving energy efficiency and reducing power demand. The first column lists major components of the system, the second column lists mitigation techniques, and the third column shows the improvement possible. These are improvements that could be realized using a system approach to mitigate energy issues associated just with the communications and computation functions of the Land Warrior.

To make progress toward providing adequate power for soldiers on the battlefield, the Army must shift its focus from providing energy to reducing energy demand, and it must do the hard job of developing a realistic mission profile. Recommendations based on these findings are considered of overarching importance in successfully confronting the issues of soldier power.

Future Warrior Goal

The Army envisions a future uniform-and-electronics ensemble for the Future Warrior. The committee believes that soldier electronics requiring a mere 2-W average, 5-W peak power is attainable in the far term if the recommendations of this study are fully implemented. By adopting state-of-the-art commercial design practices and incorporating energy-efficient technologies, peak power demand on energy sources can be reduced, thereby increasing the combat effectiveness of individual soldiers and extending the duration of their missions.

Concepts for powering the reduced needs of future soldiers should take advantage of likely reductions in the scale and distribution of power demand and consider options such as energy-harvesting to provide reliable power sources at such low power levels.

Recommendation 11: The Army should aim for a future soldier system capable of no more than 2-W average power, 5-W peak power. Achieving this will free the soldier from worries about power shortages on the battlefield and greatly enhance combat effectiveness.

Determining Energy Needs

The surest way to manage power is to utilize power-down technology for devices with heavy duty cycles. This requires detailed knowledge of duty cycles for the components as used in soldier operations. Additionally, the power dissipation of components in standby mode should be reduced as much as possible. This will become an increasingly important issue in the future owing to increased leakage currents.

Rather than crude duty cycle guesses, actual measurements of dynamic loads are needed to enable simulations of the dynamic operation of LW electronics synchronized with a power source simulator. Given a mission scenario, a suite of soldier equipment, and the size or makeup of a combat team, the Army should be able to determine an optimum type, quantity, and distribution of power sources, as well as fuel requirements. Full simulation of OFW power sources and sinks would help to determine the directions that developments must take to have the most impact. Systems could then be designed using aggressive techniques tailored to each application and to the most likely soldier modes of inter-

TABLE ES-2 Techniques for Mitigating Energy Issues in Key Land Warrior System Components and Improvements That Could Be Realized

Component	Mitigation Technique	Improvement
Power source		
Battery	Reduce peak draw	Up to 10% more available energy
Power sink		
Communications	Energy-aware network routing	Up to 50% fewer hops, 50% less energy
	Local processing	Delineation of local versus remote processing based on communications/processing cost
Computation	Remote processing	Delineation of local versus remote processing based on communications/processing cost
	Dynamic CPU speed setting	Prediction of idle time and active power within 5% of actual

NOTE: CPU, central processing unit.
SOURCE: Adapted from Martin et al., 2003.

action, thus reducing power requirements for computation and communication by several orders of magnitude.

Simulations also have the potential to save development time and money, but they require high-performance computers and accurate system inputs. High-fidelity models based on experimental data can narrow the parameters of optimization and expedite the proper selection and matching of power sources. The Army has access to high-performance computing resources easily capable of supporting such important tasks. Ideally, the military should develop and acquire new equipment only on the basis of such models, so that the lifetime of the equipment can be maximized.

Recommendation 12: The Army should develop a modeling capability for soldier equipment that includes power sources and also enables detailed simulation, verification, and analysis of power requirements for given operational parameters.

Ensuring adequate power for soldier systems is by no means a simple problem; otherwise, the Army would not have asked the National Academies to do this study. It is a multidimensional challenge, and the solutions are found by considering not only energy sources but also energy sinks and energy management. The good news is that solutions exist in all regimes to satisfy known power requirements, and major breakthroughs in power/energy source technologies are not needed. To satisfy the needs of future warriors on the battlefield, the Army must move power to the forefront of considerations in developing and acquiring soldier electronics, especially communications. It also must invest in the means to analyze power requirements, so as to take advantage of reductions that can only be achieved by efficient power management.

REFERENCES

Martin, T.L., D.P. Siewiorek, A. Smailagic, M. Bosworth, M. Ettus, and J. Warren. 2003. A case study of a system-level approach to power-aware computing. ACM Transactions on Embedded Systems 2(3): 255-276.

NRC (National Research Council). 1997. Energy-Efficient Technologies for the Dismounted Soldier. Washington, D.C.: National Academy Press.

1

Introduction

Soldier power requirements are changing as fast as new electronics are being developed. In addition to soldier communications and computers, there are a myriad of other applications for the dismounted soldier of the future that will require portable energy, including such things as laser-designators, chemical-biological sensors, uniform ventilators, and exoskeletal enhancements.

This report assesses power/energy sources, low-power electronics, and power management technologies and provides recommendations on energy solutions for the future soldier. It also evaluates the progress that is being made toward countering increasing energy demands.

This chapter provides background information on soldier power/energy issues and the origin of the study. It presents the statement of task that was used to guide the study and the approach that was taken by the committee to complete its work. It reviews findings from other studies and workshops that are relevant to soldier power/energy issues and clarifies the characteristics of the regimes that are considered by the study.

BACKGROUND

Electronics are essential to the Army's success on the battlefield. Computers, displays, radios, sensors, and other electronics applications are keys to soldier combat effectiveness. Energy to power soldier systems, while always important, must now be viewed on a par with the other critical logistics commodities—ammunition, fuel, food, and water.

Batteries are now the mainstay of soldier-portable electronics, but the acquisition, storage, distribution, and disposal of over a hundred different battery types introduces layers of logistics management and uncertainty and adds to the risks already inherent to combat. The intense demand for batteries during Operation Iraqi Freedom, for example, exceeded manufacturing capacity, and supplies would have been exhausted if combat operations in Iraq had lasted another 30 days (Fein, 2003).

Evolution of the Land Warrior

In the early 1980s, the Army recognized that the practice of equipping dismounted soldiers with items of equipment developed discretely, without an integrated view of the overall impact on the soldier, was no longer acceptable. The concept of the soldier as a system evolved from this recognition and led to a prototype for the first Land Warrior (LW) system, which was described in a previous NRC report, *Energy-Efficient Technologies for the Dismounted Soldier* (NRC, 1997), which will be referred to throughout this report as *Energy-Efficient Technologies.*

Dismounted soldiers act as both sensors and shooters, and the Land Warrior suite of electronics is intended to improve combat effectiveness by giving them increased situational awareness. Night-vision and infrared sights extend the reach of personal weapons, computer displays provide maps and locations of friendly and enemy troops, communications send and receive information on prospective targets as well as available sources of fire beyond rifle range. Suitably equipped soldiers can relay details about local targets and bring to bear virtually unlimited firepower, a capability that would have been inconceivable as recently as the first Gulf War.

The needs for electronics that use less power and for improved power sources are further driven by the fact that the Army is undergoing a major battlefield communications revolution with the transition from platform-centric warfare to network-centric warfare. This new paradigm calls for the vast amounts of information available from many and various battlefield sensors (including the soldier as a sensor) to be sent directly to an overall battlefield network rather than just to another platform (soldier, vehicle, plane, tank, etc.). The information is thus instantly available to all battlefield elements. This possible order-of-magnitude increase in information transfer, as well as the greater use of soldier-carried sensors, could easily increase the power needs of the future LW.

The LW ensemble for dismounted soldiers is designed to satisfy requirements for regular infantry, special operations forces (SOF), and Rangers, as well as airborne, air assault, and mechanized infantry. To account for different mission requirements, the Army is also developing separate ensembles for mounted soldiers, such as the Air Warrior system for helicopter crews.

But these capabilities come at a cost. Even without LW, the soldier's physical load can exceed 100 pounds for certain missions. The weight of the ensemble may add 30 pounds or more, not counting extra batteries that might be needed to guarantee power for the mission.

Objective Force Warrior-Advanced Technology Demonstration

The Objective Force Warrior-Advanced Technology Demonstration (OFW-ATD) program will integrate LW electronics using advanced concepts and demonstrate an OFW prototype in 2004.[1] Technologies and concepts demonstrated may then serve as the basis for a future generation of LW, referred to as the Land Warrior-Advanced Capability (LW-AC), to be fielded in 2007. Land Warrior-Stryker Interoperable (LW-SI) will be the first version of the LW ensemble fielded to an Army unit of soldiers. The LW acquisition program and the selection of a lead technology integrator (LTI) for the OFW-ATD in 2003 now provide the Army with an opportunity to make improvements on the design of LW from a system perspective. At least initially, the LTI focus is on relatively near-term technologies that can be used for LW-AC, but the intention is to provide a longer-term means for upgrading LW capabilities by the successive demonstration and insertion of new technology.

The Army Program Executive Office-Soldier (PEO), responsible for both the LW-SI and OFW-ATD programs, greatly assisted the study by providing access to the LTI as well as information on soldier power requirements and issues.

Relevant Studies and a Workshop

While batteries are clearly the best solution for many soldier applications, Army research has focused on technology alternatives that might reduce soldier dependence on batteries. The Army Research Laboratory/Army Communications Electronics Command (ARL/CECOM) workshop held in October 2002 reviewed known power/energy solutions and determined that there are multiple technology solutions depending on the specific technical requirements. The solutions most relevant to future Army applications reside in three separate regimes: 20-W average with 50-W peak; 100-W average with 200-W peak; and, 1- to 5-kW high-power-draw applications (Green et al., 2002).

The 2-day workshop recommended that the Army focus on developing specific power sources for the near to mid-term and the mid- to long term. The near- to mid-term focus should be on rechargeable/disposable batteries for short missions and battery-battery hybrids for long missions. The mid- to long-term focus should be on multiple-technology hybrid systems in which a "battery" is the key component.

In addition to the ARL/CECOM workshop, three studies were conducted to analyze aspects of the growing problem of soldier power/energy sufficiency. Along with the workshop, these studies provided points of departure for the committee's work.

Energy-Efficient Technologies for the Dismounted Soldier

The NRC Committee on Electric Power for the Dismounted Soldier completed a study on soldier power in 1997 (NRC, 1997). The resulting report, *Energy-Efficient Technologies for the Dismounted Soldier*, assessed technologies in all areas and contained five overall conclusions:

- Lack of power will limit the combat effectiveness of dismounted soldiers.
- Both fueled power/energy systems and energy-efficient designs will be necessary to achieve energy sufficiency on the battlefield.
- Access to the commercial electronics world must be improved.
- Power for wireless transmissions will dominate energy demand.
- Research should be conducted in multiple areas: including advanced fuel cells, microturbines, and thermophotovoltaic converters, for the far future.

The Objective Force Soldier/Soldier Team

The Army Science Board (ASB) completed a summer study *The Objective Force Soldier/Soldier Team*, in November 2001, which included several relevant findings on soldier power (ASB, 2001). The mission of the study's panel on power was to identify, assess, and recommend advanced power system technologies for the soldier system of the future. The panel concluded that power management affords the highest payoff and is the critical technology for enabling increases in mission duration comparable to what has been achieved in commercial systems: 2× improvement would be achieved by careful implementation of software to manage existing subsystems (e.g., power on/off devices) and by

[1] Since completion of the study, the term "Future Force" has been adopted for use by the Army in place of "Objective Force."

Army energy conservation and signature management; 5× to 10× improvements would be realized by considering power and power management in the design cycle.

The ASB study recommended a power source roadmap leading from enhanced disposable/rechargeable batteries in the near term (2004), to rechargeable batteries (better than 2× improvement in power management) in the mid-term (2007), to a hybrid power system (rechargeable battery with a wearable, refuelable, and disposable source) in the far term (2012).

Portable Energy for the Dismounted Soldier

The JASONs completed a study, *Portable Energy for the Dismounted Soldier*, for the Office of Defense Research and Engineering in 2003 (JASON, 2003). Among other things, its charter was to assess fuel cell technologies and to provide insights on whether alternative energy generation technologies would be more appropriate for investment. Findings included the following:

- Several technologies have legitimate potential, at 20 W for long missions, to significantly outperform existing battery packs. All such systems are hybrids with secondary batteries or electrochemical capacitors. Existing hybrid battery/battery systems can significantly reduce soldier battery pack mass (from 20 kg to 6 or 7 kg) for certain missions.
- Engineering considerations, as opposed to fundamental physical constraints, dictate the performance of fielded systems. The application space is unique to the military.
- PEM/H_2 fuel cells (with tankage for pressurized hydrogen gas) can provide significant improvement over current primary batteries. Direct methanol fuel cell (DMFC) systems look especially promising for this application, having demonstrated output energy densities from fuel 10 times greater than current batteries.
- Microdiesel engines producing 100 to 500 W seem well suited for rapid multibattery charging using JP-8. Engineering trade-offs become severe as overall system volume and mass decrease or as power capacity per unit increases.

Past Study Efforts

The three studies and the ARL/CECOM workshop had different task statements (terms of reference) and varied significantly in depth. Each study effort included evaluations of some technologies to be used for soldier power, and more than one assessed the state of the art. The ARL/CECOM workshop was only 2 days long, but benefited from many of the findings of the studies.

Table 1-1 provides a quick overview of technologies that were considered in past study efforts. The committee used information in Appendix C of *Energy-Efficient Technologies* (NRC, 1997) and assessments of technology readiness documented in Appendix D of the present study as the bases for its findings in this report.

Statement of Task

The ARL/CECOM Energy and Power Workshop for the Soldier held on 15-17 October 2002 provided the foundation for this study effort. It described many of the relevant applications of soldier electronics and categorized energy demand in distinct regimes. As a result of the workshop, the Army

TABLE 1-1 Consideration of Relevant Technologies by Previous Studies, the Workshop, and the Present Study

Technology	NRC 1997 Study	ASB Study	JASONs Study	ARL/CECOM Workshop	Present Study
Primary battery	C	C	C	C	C
Rechargeable battery	C	C	C	C	C
Fuel cells (hydrogen)	C	C	C	C	C
Fuel cells (methanol)	NC	C	C	C	C
Fuel cells (solid oxide)	NC	C	C	C	C
Internal combustion	C	C	C	C	C
Microturbines	C	C	C	C	C
Stirling	C	NC	NC	C	C
MEMS-based electrochemical capacitors (ECCs)	C	NC	NC	C	C
Thermoelectric	C	NC	NC	C	NC
Thermophotovoltaics (TPV)	C	C	C	C	NC
Nuclear isotope	C	C	NC	C	NC
Alkali metal thermal-to-electric converters (AMTEC)	C	NC	C	C	NC
Energy harvesting; solar	C	C	NC	C	NC

NOTE: C, considered; NC, not considered.

approved a statement of task for a study to be implemented by the National Research Council Board on Army Science and Technology.

The Assistant Secretary of the Army (Acquisition, Logistics, and Technology) requested the National Research Council to determine suitable alternatives for powering future soldiers on the battlefield by accomplishing a study on portable power sources, power management, and low-power electronics technologies as follows:

1. Expand upon the conclusions from the ARL/CECOM Energy and Power Workshop for the Soldier, held on 15-17 October 2002, through the specification of both impact and feasibility of incorporating power management components, techniques and procedures for powering low-power electronic devices. The specific regimes from the workshop were: 20-watt average with 50-watt peak and 100-watt average with a 200-watt peak for up to 72-hour missions. Address power for high-power draw applications such as exoskeleton applications (1-5 kW average).
2. Assess electric power technologies to support soldier applications associated with future power and energy demands on the battlefield, e.g. expected OFW operational capabilities for the 2005-2025 timeframe, with emphasis on alternative compact high-power and energy-dense sources, power management and distribution techniques, and low-power electronics such as asynchronous microchips, smart dust, etc. Assess technical risks and feasibility associated with each of the technologies and make recommendations pertaining to their potential efficacy and utility within the context of future OFW operational capabilities. Consider risks associated with technology development, integration of hybrid generators and sources, adaptation of commercial technologies, and battlefield logistics. Systems concepts involving appropriate power sources, power management and low-power electronics are to be specified and delineated.
3. Update the technologies evaluated in the 1997 NRC report on Energy-Efficient Technologies for the Dismounted Soldier including changes in individual technology development trends. Determine advantages and disadvantages for appropriate technologies in prospective application areas. Develop standard measures to facilitate comparison.
4. Prepare a consensus report documenting the study results and containing findings and recommendations to assist the Army in its development program. Prioritize the energy source alternatives appropriate to each application. Propose S&T objectives leading to the future incorporation in the Objective Force Warrior program. The report will include:
    a. Recommendations for examined technologies with high benefit for target regimes with detailed justification for technology selection or rejection.
    b. Recommendations for power distribution techniques for soldier systems. Applicability of low-power electronics, such as asynchronous microchips, smart dust, etc., to soldier device loads.
    c. Recommendations for centralized vs. distributed power management for soldier systems including software/hardware techniques for control and conversion.
    d. Applicability of examined technologies to single type sources vs. hybrid sources considering logistics, versatility, utility, environmental factors, safety, reliability, logistic infrastructure, manufacturability and availability.
    e. Recommendation for recharging from soldier carried sources, robots (or vehicle) or fixed platforms.
    f. Recommendations for predictive models and modeling techniques that would elucidate power use and management.

STUDY APPROACH

The statement of task contained multiple tasks requiring specific areas of expertise to ensure their accomplishment. As a result, committee members were selected who had expertise in the relevant technologies, including primary and rechargeable batteries, fuel cells, electrochemical devices and systems, small engines, hybrid systems, and low-power electronics and design, as well as in military logistics and operations.

The chair determined that to accomplish its primary task, the study should rely on the expertise of the members to make realistic assessments of the possible solutions in each of the regimes and to focus on technologies that can enable systems within the near, medium, and far terms. Technologies that can enable viable power/energy systems were then compared and ranked. The assessments also provided the basis for identifying suitable Army research objectives.

The committee evaluated Army progress toward resolving soldier power issues by reviewing the LW-SI acquisition program and the OFW-ATD. It received briefings on anticipated soldier applications in the higher power regimes from both the Army and the Defense Advanced Research Projects Agency (DARPA).

The committee assessed advances in low-power electronics and investigated applicable areas of power distribution and management. It updated earlier assessments of trends in commercial electronics contained in NRC (1997) and developed future warrior design concepts. It then reached consensus on its specific recommendations for the Army.

REPORT ORGANIZATION

This report documents the study findings and recommendations and is organized in accord with the task statement and the study approach described above. The report is organized in chapters as follows.

Chapter 1 (Introduction) provides background information and the statement of task for the study. Chapter 2 (Technology Alternatives) describes the most realistic power/energy technology solutions in each of the three regimes of prime concern for present and future Army applications. Chapter 3 (Power System Design) discusses guidelines for the efficient integration of power source technology. Chapter 4 (Soldier Energy Sinks) discusses the range of power demands of soldier applications and the key role played by logistics in determining the viability of energy solutions.

Chapter 5 (Progress) summarizes committee observations on the progress made by the Army since the 1997 NRC study report and on recent commercial trends in technology. Chapter 6 (Future Warrior Design Concepts) discusses promising approaches to design and integration of future soldier systems, barriers to implementation of energy solutions for soldier systems, and the impact of user interaction on power demand. Chapter 7 (Recommendations) summarizes the findings and presents the study recommendations. Appendix C (Measures of Performance) describes the methods used to classify the power sources and provides means for evaluating new power sources. Finally, Appendix D (Source Technologies) describes the characteristics of the power sources considered and reports on advances in technology since *Energy Efficient Technologies* (NRC, 1997).

2

Technology Alternatives

This chapter discusses the results of the committee's assessments of power source options for the three power regimes. It describes the assumptions about each regime, key terms and metrics, and standards for the selection of appropriate power source technologies in the context of 24- and 72-hr missions. The chapter concludes with committee findings on appropriate science and technology (S&T) objectives for the Army to undertake for the near, medium, and far terms.

Detailed discussion of all technologies covered in this chapter, including the factors considered in assigning technology readiness levels (TRLs), sources, and references, are contained in Appendix D (Source Technologies). Appendix C defines measures of performance and other terms.

The NRC report *Energy-Efficient Technologies for the Dismounted Soldier* (NRC, 1997) described and assessed most of the energy sources considered in the current study; readers who are not familiar with basic battery, fuel cell, and hybrid technologies are encouraged to use it as a reference. Table 2-1 provides an overview of all power source technologies considered in the current study and notes the technologies that have emerged since the earlier report was published.

This chapter describes and compares power source alternatives for three power regimes for missions up to 72 hours. The regimes are 20-W average, 50-W peak; 100-W average, 200-W peak; and 1- to 5-kW average. Requirements data on all missions possible for each regime do not exist, so assumptions were made regarding likely applications and power demands in each regime. These assumptions enable power source options from the ARL/CECOM workshop to be compared with options and conclusions from the present study.

The statement of task specifies average power requirements and periodic peak power requirements but not duty cycles. Since duty cycle will significantly affect the choice of individual or hybrid power sources, the committee makes quantitative comparisons between power source alternatives for the average loads mentioned above. Power sources that can accommodate the peak loads are discussed, but no quantitative comparisons are possible without duty-cycle details. Chapter 3 addresses in detail how peak loads can be handled and discusses their impact on the overall power system.

ASSUMPTIONS

The committee made certain assumptions about the electronics applications included within the three regimes called out in the statement of task. These assumptions are consistent with the baseline ARL/CECOM Workshop.

The 20-W regime was assumed to cover power solutions for computers, radios, sensors, displays, and especially, the electronics subsystems of the Land Warrior (LW) ensemble. Table 2-2 lists devices now being considered for inclusion in the Objective Force Warrior-Advanced Technology Demonstration (OFW-ATD). The list does not include an evaporative cooler estimated to demand 10 W, which is assumed to require a separate power source in the near term.

Power solutions in this regime, which currently uses logistics-intensive and costly disposable batteries, are of primary importance to the Army. The Army will continue to support a wide variety of electronics in this regime, and new power system solutions are urgently needed to enhance both logistics and combat effectiveness.

The single application envisioned at the ARL/CECOM Workshop for the 100-W regime was a portable recharger for rechargeable batteries. The committee also considered solutions for the laser target designator—current versions of which demand an average 100 W (peak power up to 180 W)—and microclimate cooling to be in this regime. Finally, the committee assumed that the 1- to 5-kW regime includes power-intensive capabilities, such as the exoskeleton.

TABLE 2-1 Overview of All Power Source Alternatives

Power System	State of the Art, 1997[a]	State of the Art, 2003	Item Considered	Scaling Laws	Impact on Soldier Power
Primary battery (includes metal/air)	Mature. Up to 800 Wh/kg in low-specific-power configurations	Mature. SOA not significantly advanced beyond NRC (1997) report.	Energy density. Safety. Power density. Environmental impact.	Known	Heavy, one-time use. Current battery of choice for combat missions. Potential for use in hybrids.
Secondary battery	Mature. Li ion: 100 Wh/kg in development.	Mature in commercial applications. Li ion: 140 Wh/kg available; 200 Wh/kg in development.	Energy density. Cycle life. Power density. Safety and cost.	Known	Stand-alone energy supply for many missions. Can be used in hybrid mode for high-energy missions.
Fuel cell (hydrogen)	Exploratory development. Many systems at laboratory scale. Power levels to 150 W considered.	Beta prototypes with various hydrogen sources tested in field. Power to 150 W.	Fuel reformers. Water management. Safety.	Known	New capability; potential for use in hybrid system. Less weight. Cost savings. Requires new battlefield fuel.
Fuel cell (methanol)	Emerging. Not considered.	Beta prototypes developed at power levels of 20 to 50 W. 20% efficiency.	Fuel and fuel crossover. Catalyst. Cost.	Known	New capability. Less weight. Cost savings. Requires new battlefield fuel.
Fuel cell (solid oxide)	Emerging. Not considered.	Emphasis on small sizes. Laboratory prototypes in 20-W range. Research in high-capacity designs.	High temperature. Materials. Integration and systems.	Known	New capability. Less weight. Easier to utilize battlefield fuels. More efficient.
Internal combustion	Some versions mature. Hobby application sizes coupled to generators. No commercial products on market.	Commercial applications with motor-alternator combinations in 30 to 100 W/kg range. Efficiencies greater than 20% in 500-W sizes. Emerging modified hobby engines operate on diesel.	Fuels. Vibrations. Life.	Known	Inexpensive technology. Potential for high-energy missions. Can probably be made to function with JP fuels. Current role as battery charger.
External combustion (includes Stirling)	Not considered.	100 W/kg specific power demonstrated for motor-alternator with efficiency of 29%. System efficiencies projected to be >20%. Laboratory 35- to 50-W systems available for beta prototypes; 1- to 2-kW beta prototypes available with ~20% system efficiencies. System-specific power appears to be around 30 W/kg.	Fuels. Specific power. System-specific energy. Signatures.	Known	New stealth capability. Inexpensive technology. Can be made to operate on JP fuels. Potential for high-energy missions.
Microturbine	Emerging. Considered promising.	Not considered owing to lack of progress in producing workable systems.	Fuels. Specific power. System-specific energy. Materials. Cost.	Unknown	

continued

TABLE 2-1 Continued

Power System	State of the Art, 1997[a]	State of the Art, 2003	Item Considered	Scaling Laws	Impact on Soldier Power
Thermoelectric	Some versions mature. Low potential. Best system efficiency on order of 5%; converter efficiencies projected to 10%.	Insufficient progress to consider for current applications. Progress in new high-ZT materials makes technology worth watching for long term.	Efficiency. Materials-specific power. System-specific energy.	Known	Not applicable owing to low efficiency. Possible niche application in small sizes.
Thermo-photovoltaic (TPV)	20% TPV cells demonstrated. System projections to 20%.	Not considered owing to lack of progress in systems.		Known	
Nuclear isotope	Limited consideration. Rejected owing to cost, safety, environmental considerations, and lack of infrastructure.	Not considered.	Safety. Environmental impact. Cost. Public acceptance.	Known	
Alkali metal thermal-to-electric converter	Speculative technology. Systems projection to 500 W/kg.	Not considered owing to lack of progress.		Known	
Energy harvesting; solar	Some versions mature.	Considered for low-capacity niche applications.		Known	Driver for reducing power demand.

NOTE: SOA, state of the art; Li ion, lithium ion; JP, jet propellant; ZT, thermoelectric figure of merit.
[a]NRC, 1997.

FIGURES OF MERIT

The governing figure of merit used to discriminate among and, in the final analysis, to rank-order the technologies was total system mass (as estimated from the specific energy of the underlying technology). Other figures of merit used to evaluate the technologies and systems are described in detail in Appendix C. The committee estimated technology readiness levels (TRLs) to determine the systems worthy of consideration. Definitions for the nine TRLs are also contained in Appendix C.

It is important to note that technical figures of merit for many of the emerging power sources were not available, and in a few instances, the information was not considered reliable enough. Many of the technologies evaluated were in various states of development, and the committee made some assumptions about expected system characteristics and performance. These assumptions and/or extrapolations are documented so that the reader can better judge the relative merits and risks of the technology options.

POWER SOURCE SOLUTIONS

Batteries represent the ideal solution for soldier power and energy applications. Only when every effort has been made to conserve and manage energy and it is found that batteries cannot meet requirements should air-breathing systems, such as fuel cells or small engines, be considered. A heavy price is paid when these nonbattery options are used, including the requirement for continuous airflows, sensitivity to contaminants, temperature restrictions, possible orientation dependence, acoustic and thermal signatures, nonstandard fuels, surface and exhaust temperature, and exhaust gas contamination. That being said, there are mission requirements today for soldiers that exceed the reasonable capabilities of battery technology, and in these cases air-breathing alternatives are emerging to meet these needs.

The rationale used to compare alternatives provided a framework for selecting power source options for the different power/energy regimes. The analyses considered all that is presently known about existing and emerging power source performance, and the technologies selected for further consideration by the Army are those that will most likely meet mission requirements with respect to specific power and specific energy. It is always possible that for particular missions, other factors, such as acoustic and thermal signature, operating temperature, fuel, orientation dependence, logistics, etc. will be more important than the specific power and specific energy of the system. Ultimately, the Army will

TABLE 2-2 Devices in 20-W Regime Planned for Objective Force Warrior (OFW)-Advanced Technology Demonstration

Function	Power Demand[a] (W)
Communications	
Soldier radio	7.8
Squad radio	6.2
UAW/robotic vehicle	6
Computer displays	
Handheld flat panel	7.05
Helmet-mounted	0.5
Integrated sight—module display	3
Sensors	9.5
Computer	17.42
Total	57.97

[a]Breakdown of OFW numbers:

- Soldier radio, 7.80 W (JTRS numbers are not available; assumed the same as Stryker MBITR radio);
- Squad radio, 4.40 W (communications processor card) + 0.60 W (WLAN card) + 0.60 W (VoIP processor) + 0.60 W (WLAN antenna);
- UAW/robotic vehicle, 3 to 10 W for como-crypto interface (Brower, 2003);
- Handheld flat panel, 6.30 W + 0.75 W (handheld keyboard and cable);
- Helmet-mounted display, 0.50 W;
- Integrated sight display, 3 W (HIA module including breakaway connection to body PAN);
- Sensors, 2.15 W (thermal weapons sight) + 1.10 W (daylight video sight) + 4.00 W (multifunction laser) + 1.50 W (GPS) + 0.25 W (dead reckoning module) + 0.50 W (microphone/speaker assembly); and
- Computer, 2.10 W (computer assembly) + 10.9 W (computer processing card) + 3.42 W (PAN body hub) + 1.00 W (PAN weapon hub).

NOTE: UAW, universal access workstation; JTRS, Joint Tactical Radio System; MBITR, multiband intra/inter team radio; WLAN, wireless local area network; VoIP, Voice over Internet Protocol; HIA, high integration actuator; PAN, primary area network; and GPS, Global Positioning System.

SOURCE: Adapted from Erb, 2003, and Brower, 2003.

need to make the final decision based on trade-offs to suit specific mission requirements.

The statement of task includes peak power requirements for two of the three power regimes of interest. How peak power is handled by the energy source will depend on the duty cycle. In general, hybrid systems can enable high-efficiency operation over an entire power spectrum of operation provided that the requirement for a separate peak power source warrants the additional weight and volume. If a separate battery is chosen to meet the minimum and peak power demands, it must be capable of delivering the desired power and part of the total energy. The energy converter portion of the hybrid must provide the average power and all of the balance of the total energy. This includes energy sufficient to fully recharge the battery during the nonpeak or low-power operating portion of the duty cycle.

There are three important issues that need to be addressed when making comparisons between figures of merit for the various power sources. The first is encountered when energy storage and energy conversion devices, e.g., batteries and fuel cells, are to be compared. Reasonable comparisons can be made if, and only if, the total energy content, including converter, fuel, and fuel tank, of the energy conversion device is compared with an energy storage device, such as a battery pack, having an equal amount of stored energy. In other words, the total energy produced by each system must be measured under the identical load conditions (power profile) to obtain an accurate comparison between the two.

A second issue is related to technology maturity. For many emerging technologies, fully packaged systems are not available. The system dry weight, including the fuel tank, the quantity of fuel, the energy content of the fuel, and the energy conversion efficiency, are all needed to compute performance metrics. Efficiency data are available for some emerging technologies. If the quantity and energy content of the fuel are known, all that is needed is the dry weight of the optimized system for a specific mission requirement. Reasonable estimates of system dry weights can be inferred from breadboards, brassboards, prototypes, and commercial products. In the latter case, allowances need to be made to account for differences between commercial and military priorities, e.g., the weight may be unimportant to a commercial customer whereas it is critical to the soldier. Hence, the commercial product is not optimized for weight, and the specific energy of the technology may be underestimated. To make meaningful comparisons among alternatives, energy conversion system dry weights were estimated based on assumptions that are explained in Appendix D with references.

The third issue relates to battery comparisons. Performance specifications for batteries are given for specific cell sizes and discharge rates. The specific energy data quoted in this study are valid for the discharge rates under consideration. However, there will be a packaging penalty (weight and volume) for battery packs. For example, a lithium ion laptop computer battery may be in a square configuration 2.5-cm thick, but in reality there are eight cylindrical 18650-Li ion cells inside this package. The performance specifications of the 18650-cell should be discounted to account for this. A good rule of thumb is to deduct 15 percent from the cell performance figures.

Fueled systems, which are in various stages of development, can be used to replace batteries or supplement them as part of a hybrid system. For systems supplying more than about 1 kWh, fueled systems offer a significant mass advantage over batteries. Figure 2-1, taken from the 1997 report, illustrates this point. It can be seen that the battery mass is directly proportional to mission energy requirement. In contrast, for fueled systems, mass comprises the fuel (including fuel tank) mass, which is a function of the mission energy requirement, and the energy converter mass, which is a function not of the mission energy requirement but of the mission power requirement. The *y* intercept in the figure is the dry mass of the energy converter and the slope is the product of the energy content of the fuel and the system energy conversion efficiency. These issues are explained in detail in the earlier report (NRC, 1997).

ANALYSIS OF ALTERNATIVES

The known performance of state-of-the-art lithium/manganese dioxide and lithium/carbon monofluoride (Li/MnO_2 and Li/$(CF)_x$) primary batteries and lithium ion (Li ion) rechargeable battery technologies was compared with that of promising energy conversion technologies. The interpretation of data for new technologies was intentionally conservative, with every effort made to use performance data obtained from completely packaged systems. In some cases, projections were made from subsystem data if system data were unavailable. Assumptions and references used to make these projections are documented in Appendix D. Low-TRL concepts (e.g., lithium/air, carbon/air) were not compared if too many assumptions were needed to predict system-level performance. Similarly, low-performance technologies—that is, those that were known not to exceed lithium battery performance—were not included in the analyses.

Plots of total system mass including fuel versus 24- and 72-hr mission durations were developed for alternatives in each power regime (Figures 2-2 to 2-5). The corresponding numerical data are included in Tables 2-3 and 2-4. The technologies were rated on their ability to provide both average and peak powers for a given regime. The battery mass needed to produce the equivalent amount of energy was calculated from cell data. Because the typical 15 percent mass penalty for packaging these cells into battery packs was not included, battery performance is slightly overestimated in these charts

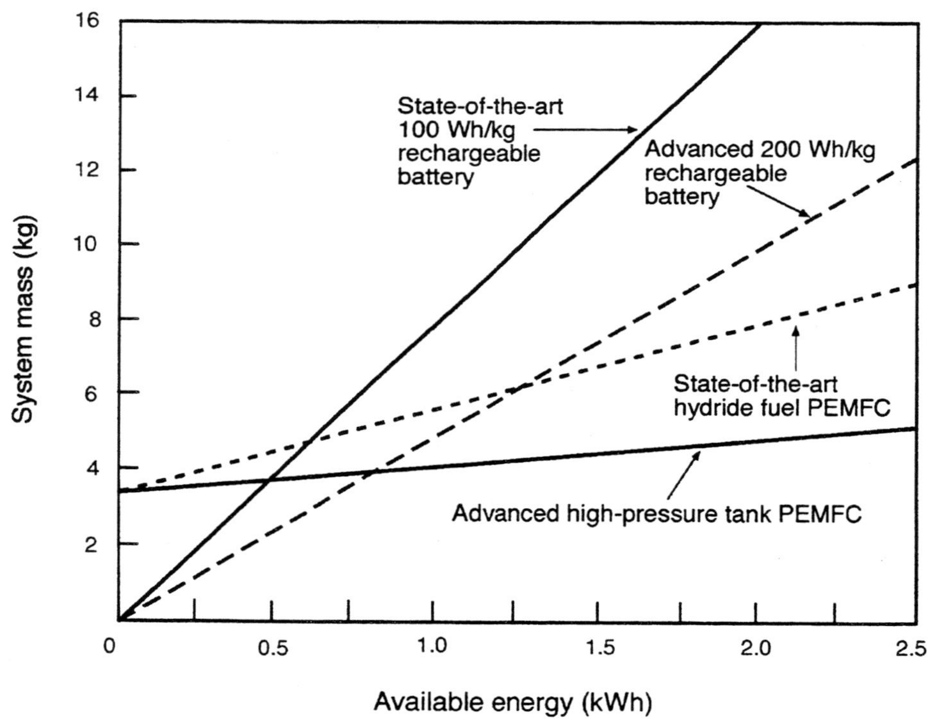

FIGURE 2-1 Graph showing the crossover points for battery and fuel cell power systems as functions of available energy and system mass. The assumed system power level is 5 W. PEMFC is proton exchange membrane fuel cell. SOURCE: NRC, 1997.

TABLE 2-3 Comparison of Soldier Power/Energy Sources for 20-W Average Power Missions of 24 and 72 Hours

Technology	Mission Length (hr)	System Mass (kg)	Power (W)	Total Energy (Wh)	Specific Energy (Wh/kg)	TRL
Stirling (JP-8)	24	0.82	20	480	588	2
PEM/H_2 (5,000 psi)	24	1.15	30	720	626	7
PEM/$NaBH_4$	24	1.16	20	480	414	6
Li/$(CF)_x$ (SOA, primary)	24	1.54	20	946	614	8
Li/MnO_2 (SOA, primary)	24	1.71	20	480	280	8
IC (JP-8)	24	2.00	50	1,200	600	4
DMFC	24	2.04	20	480	235	6
Li/MnO_2 (LW, primary)	24	2.46	20	480	195	9
SOFC (butane)	24	2.68	20	480	179	4
Li ion (SOA, rechargeable)	24	2.82	20	480	170	9
Li ion (LW, rechargeable)	24	3.31	20	480	145	9
Stirling (JP-8)	72	1.20	20	1,440	1,200	2
PEM/H_2 (5,000 psi)	72	2.09	30	2,160	1,033	7
Li/$(CF)_x$ (SOA, primary)	72	2.35	20	1,440	614	9
PEM/$NaBH_4$	72	2.59	20	1,440	556	6
IC (JP-8)	72	3.00	50	3,600	1,200	4
DMFC	72	3.01	20	1,440	478	6
SOFC (butane)	72	3.90	20	1,440	369	4
Li/MnO_2 (SOA, primary)	72	5.14	20	1,440	280	8
Li/MnO_2 (LW, primary)	72	7.37	20	1,440	195	9
Li ion (SOA, rechargeable)	72	8.47	20	1,440	170	8
Li ion (LW, rechargeable)	72	9.94	20	1,440	145	9

NOTES: Table is sorted by system mass. TRL, technology readiness level; JP-8, jet propellant 8; PEM, proton exchange membrane; SOA, state of the art; IC, internal combustion; DMFC, direct methanol fuel cell; LW, Land Warrior; SOFC, solid oxide fuel cell.

TABLE 2-4 Comparison of Soldier Power/Energy Sources for 100-W Average Power Missions of 24 and 72 Hours

Technology	Mission Length (hr)	System Mass (kg)	Power (W)	Total Energy (Wh)	Specific Energy (Wh/kg)	TRL
IC (JP-8)	24	2.70	100	2,400	889	4
Stirling (JP-8)	24	4.00	100	2,400	600	2
SOFC (JP-8)	24	4.24	150	3,600	849	4
PEM/H_2 6%	24	6.21	100	2,400	386	7
Li/$(CF)_x$ (SOA)	24	7.69	100	2,400	614	8
Li/MnO_2 (SOA)	24	8.57	100	2,400	280	8
DMFC	24	11.40	150	3,600	316	6
Li/MnO_2 (LW)	24	12.31	100	2,400	195	9
Li ion (SOA)	24	14.12	100	2,400	170	8
Li ion (LW)	24	16.55	100	2,400	145	9
IC (JP-8)	72	5.70	100	7,200	1,263	4
Stirling (JP-8)	72	6.00	100	7,200	1,200	2
SOFC (JP-8)	72	7.42	150	10,800	1,456	4
PEM/H_2 6%	72	10.93	100	7,200	659	7
SOA Li/$(CF)_x$	72	11.70	100	7,200	614	8
DMFC	72	18.60	150	10,800	581	6
Li/MnO_2 (SOA)	72	25.71	100	7,200	280	8
Li/MnO_2 (LW)	72	36.92	100	7,200	195	9
Li ion (SOA)	72	42.35	100	7,200	170	8
Li ion (LW)	72	49.66	100	7,200	145	9

NOTES: Table is sorted by system mass. TRL, technology readiness level; JP-8, jet propellant 8; IC, internal combustion; SOFC, solid oxide fuel cell; PEM, proton exchange membrane; SOA, state of the art; DMFC, direct methanol fuel cell; LW, Land Warrior.

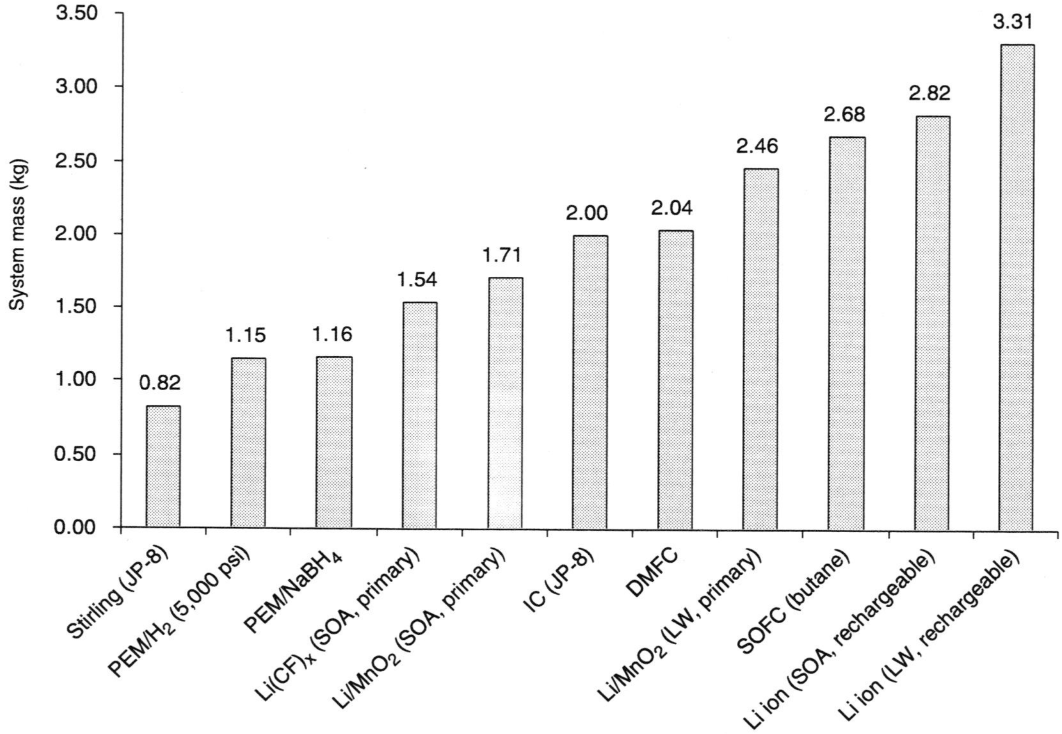

FIGURE 2-2 24-hr mission at 20-W average power.

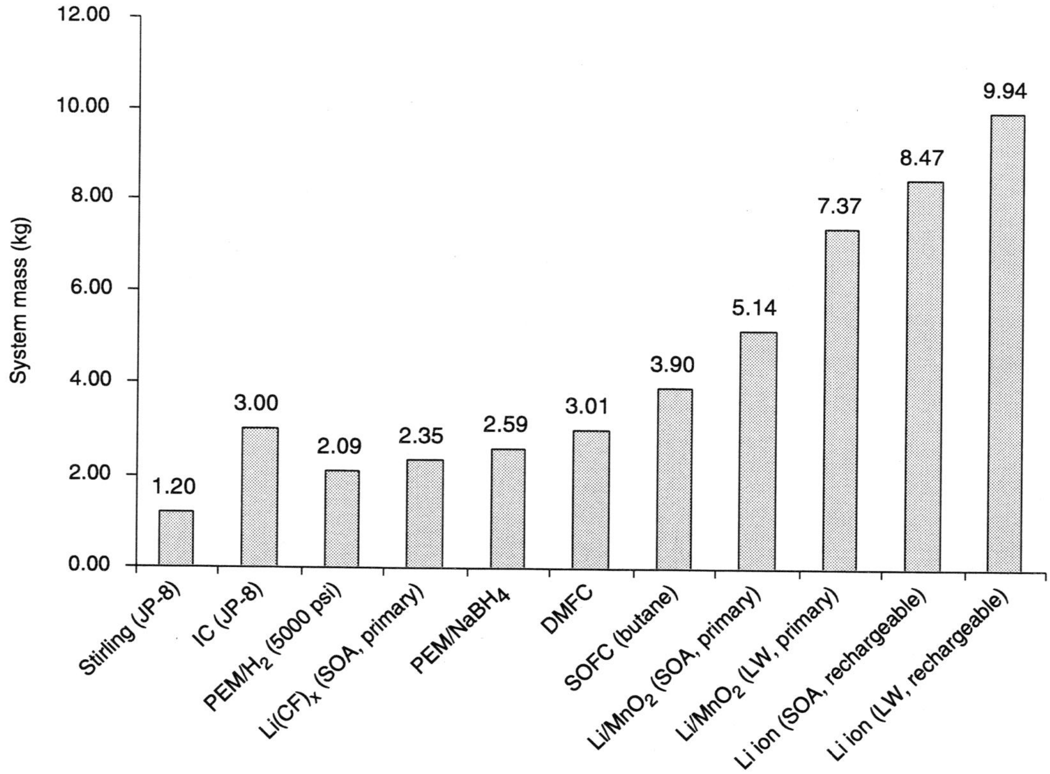

FIGURE 2-3 72-hr mission at 20-W average power.

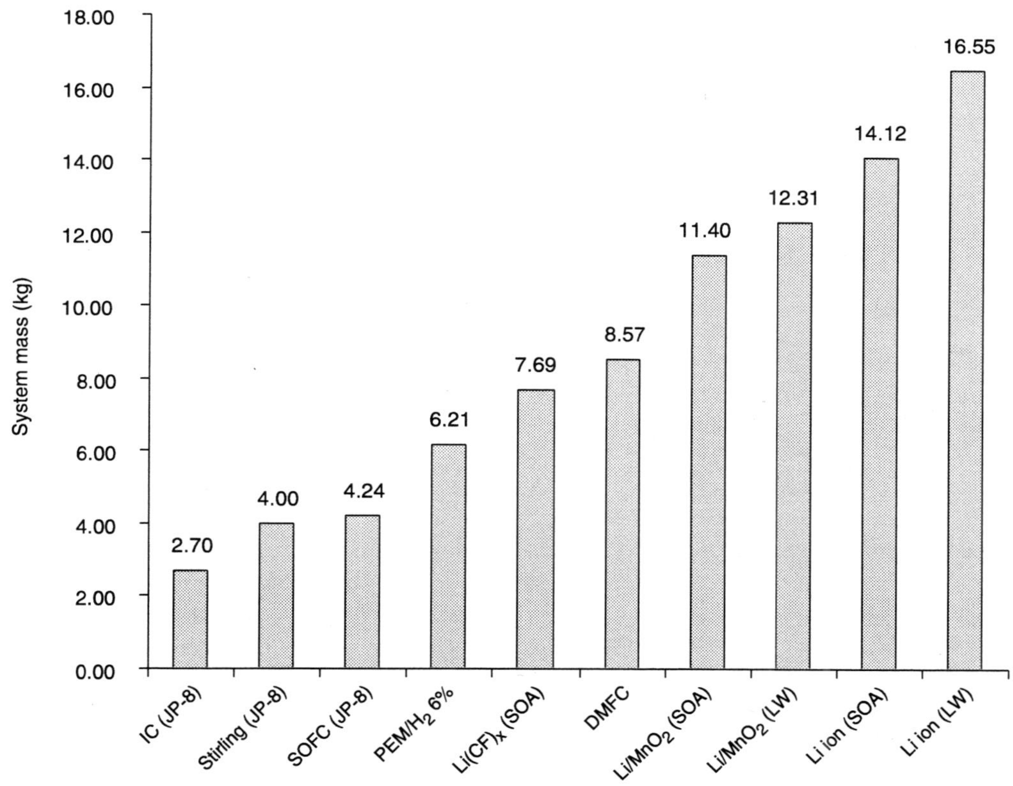

FIGURE 2-4 24-hr mission at 100-W average power.

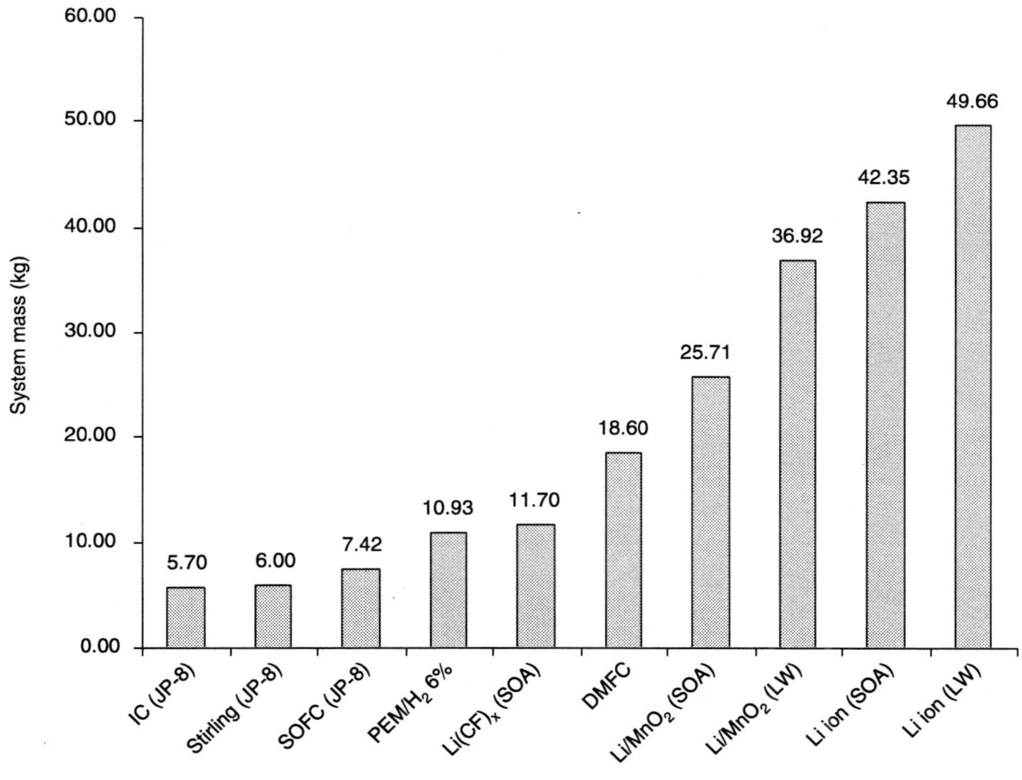

FIGURE 2-5 72-hr mission at 100-W average power.

for the state-of-the-art primary Li/MnO_2 and $Li/(CF)_x$ and rechargeable Li ion batteries. The packaged LW batteries (Li/MnO_2 and Li ion) are included as examples of the state of the art in the tables and charts.

The committee determined science and technology objectives for Army investment in the near term (2010), medium term (2015), and far term (beyond 2015) based on the technologies selected and overall results of its comparison of alternatives in each regime. Far-term objectives are also discussed in Chapter 6. Energy per unit of system mass, i.e., specific energy, served as the primary basis for selecting technologies for the Army to pursue. To be considered a viable alternative, a technology had to have demonstrated a TRL that would allow the committee to estimate its performance in a power/energy source system. Because of this, the Army will need to conduct detailed trade studies (e.g., specific energy versus logistics, signature, cost) to determine if a selected power source technology is actually suited for a particular application and/or mission. The technical characteristics of the technologies that are documented in Appendix D to this report should facilitate such trade studies. Ideally, simulated models that incorporate equipment inventory, load profiles, mission duration, and environmental conditions should be used to determine the best overall power solution for a given application.

20-W Average Power

Current power source development goals are listed in Table 2-5. The 12-hr goals are certainly achievable with primary batteries, but they are still a stretch for rechargeable cells. The committee believes that incremental improvements can be made in near-term programs to help rechargeable cells achieve the 12-hr mission goals. On the other hand, there is no battery capable of meeting the 72-hr mission goal.

Batteries have shown a continuous, steady increase in energy density for the last 40 years, from about 30 to 300 Wh/kg. The possibilities for further improvement are good and need to be pursued aggressively, but the pace of improvement is likely to continue to be slow compared with other areas of technology development. Thus, in order for the Army to meet its 72-hr goal in the near term, it must consider investing in both long-term, relatively high-risk programs such as Li/air and shorter-term hybrid and non-battery systems.

Total system mass versus 24- and 72-hr mission lengths is plotted for a 20-W average power mission in Figures 2-2 and 2-3. Numerical data are presented in Table 2-3. The battery technologies chosen for comparison are primary Li/MnO_2 (state-of-the-art (SOA) and LW versions), primary SOA $Li/(CF)_x$ (Eagle-Picher LCF-112, DD cell) and rechargeable lithium ion (SOA and LW versions). The data for proton exchange membrane/sodium borohydride (PEM/$NaBH_4$) (Lynntech) and direct methanol fuel cell (DMFC) (Ball) data are based on complete system demonstrations (TRL 6). The proton exchange membrane/hydrogen (PEM/H_2) data are based on a packaged 30-W system (Ball, TRL 7). The solid oxide fuel cell (SOFC) point is based on projections from breadboard testing of a butane-fueled system (AMI, TRL 4). The internal combustion (IC) (D-Star, TRL 3-4) and Stirling (Sunpower, TRL 1-2) points are projections based on demonstrated engine performance and balance-of-plant (BOP) estimates; for example, the Sunpower motor-alternator is at TRL 4, but lack of BOP technologies reduces the overall system TRL to 1-2.

Modest gains over rechargeable batteries are achievable with some energy conversion systems for the 24-hr mission. While the Stirling engine appears to be the lightest option, further development is necessary to validate this point due to the relatively low TRL level for this system technology. The most attractive candidates that have been demonstrated at relatively high TRL levels are PEM/H_2 systems. With the advent of small, lightweight stacks (Protonex), the performance could improve even further. It should be noted that the PEM/H_2 data are for a 30-W system that includes a Li ion battery, which could be used for brief peak loads. This system has the highest specific energy for the 24-hr mission when one considers the total energy delivered. If hydrogen is unacceptable, DMFC could be developed further and could outperform primary batteries. Because of their modest specific power, nine $Li/(CF)_x$ cells will be required to meet the 20-W power demand. These nine cells provide the lightest battery option and will deliver 20 W for 47 hours, nearly twice as long as the other batteries or systems considered.

Energy conversion systems become more attractive at the 72-hr mission length where there is a potential for reducing the total mass by a factor of 4 or 5. They are all

TABLE 2-5 Power Source Development Goals for Soldier Systems

Load (W) (average/peak)	Mission Time of 12 hr	Mission Time of 72 hr	Required Weight (kg)
20/50	240[a]	1,440[a]	1.0
100/200	300[a]	1,800[a]	4.0

[a]These numbers are calculated specific energy in Wh/kg.

SOURCE: Pellegrino, 2003.

significantly better than rechargeable lithium ion and, with development, will be much better than primary Li/MnO_2. A caveat is that if these systems are being used as part of an overall hybrid system (battery plus additional electronics), additional weight and volume will have to be added to the total system mass calculation. This will be the case for air-independent operation and for duty cycles that cannot be accommodated by the prime power source. Depending on the duty cycle, a lithium ion battery or capacitor can handle the 50-W peak requirement.

The PEM and DMFC demonstration data depicted in Figure 2-3 show much better performance than batteries for the 72-hr mission. There are opportunities for significant mass reduction in both of these systems, which will make these comparisons even better. It should be noted that these systems are more complex than batteries alone and require that attention be paid to such things as start-up, fuel and oxidant control, water management, shutdown, and storage below freezing. The PEM system data shown in Figures 2-2 and 2-3 represent fully packaged systems. Other hydrogen sources for PEM should be evaluated in the context of a complete system as they become available. In general, any hydrogen storage or generation concept that yields over 6 percent hydrogen storage based on the total mass of the system should be assessed for Army use. One interesting hydrogen generation alternative is ammonia cracking. Ammonia can be stored at reasonable pressures, thus reducing the mass of the storage container and valve. The reformate is a mixture of hydrogen and nitrogen and does not contain the sulfur and carbon monoxide contaminants found in hydrocarbon reformate. To yield pure hydrogen with no nitrogen dilution, ammonia reforming would be implemented with a hydrogen separation membrane, which would reject the nitrogen. These systems are under current development (MesoFuel). Recent work in small hydrocarbon reformers (Altex) shows promise for integration with small fuel cells. These systems are better suited for integration with SOFCs as the relatively high temperature of the reformate is more compatible with SOFC than with PEM stacks.

SOFCs and small engines have the advantage of operating on energy-dense hydrocarbon fuels that are readily available in the field, but SOFCs are more sensitive to contaminants in the fuel than engines. While SOFC is probably the least developed of these technologies in this power range, significant weight reductions are likely to be achieved in the future. It should be noted that the SOFC point is based on a breadboard system with a catalytic partial oxidation (CPOX) reformer. SOFC has a much lower acoustic signature than IC engines but is less rugged and will require accommodation for shock and vibration. This has been achieved for the packaged PEM and DMFC systems constructed by Ball Aerospace, and similar techniques could be applied to SOFC systems. AMI has demonstrated that full power can be obtained for a 20-W SOFC stack in less than 3 minutes, which is a major achievement for this technology. And, unlike large fuel cells (greater than 1 kW), these small versions can be stopped and started multiple times without detriment (see Appendix D).

MEMS advances are likely to play a role in the future development of fuel cell and other power systems. They will enable the miniaturization of balance-of-plant (BOP) components, including integrated fuel processors, for many of the conventionally fabricated systems now under consideration. The contribution of MEMS to BOP is more critical to soldier power solutions than its contribution as a prime generator of energy.

The mass projected for small engines considered in this regime required some speculation on BOP components and scaling laws, with more assumptions for the Stirling data than for the IC data. Note that the D-STAR engine is a 50-W device, so that while it is heavier than some of the other options, the specific energy is among the highest (Table 2-3) because more energy is produced in a given time. Both IC and Stirling engines are capable of operation on logistics fuels—that is, fuels (such as JP-8) that are readily available in bulk on the battlefield. D-STAR has made considerable progress on suppressing acoustic signature, which is a potential drawback for this technology. Stirling engines with minimal acoustic signature have been demonstrated.

Findings for 20-W Average Power Regime

- Energy conversion systems are somewhat better than batteries for the 24-hr mission but are more complex to operate. However, mission length is easily extended by fuel addition, and the longer the mission the more competitive energy conversion systems become. An added benefit is the ability to continue a mission (provided enough fuel is left) even if resupply is not forthcoming in 24 hours.
- Energy conversion systems are currently one fourth as massive as rechargeable batteries for longer missions (72+ hours).
- From a logistics perspective, prepackaged fuels can be treated as battery packs as long as appropriate safety and handling procedures are developed. Such prepackaged fuels will enhance the attractiveness of fueled energy conversion alternatives.
- Some energy conversion technologies in the 20-W power range may be capable of operating on bulk logistics fuels.
- Logistics issues are as important as performance in determining which power source to use for soldier systems.

Science and Technology Objectives

Science and technology (S&T) objectives consistent with the committee's selection of alternatives in the 20-W regime are listed below. The objectives are listed in order of importance, along with the key development issues to be resolved.

There are eight near-term objectives:

- Develop batteries for the 24-hr mission with specific energies greater than 300 Wh/kg. Key development issues: low-cost, safe routes for the synthesis of carbon monofluoride $(CF)_x$; custom electrode formulation and materials optimization to support higher rates with minimum impact on specific energy; lithium sulfur (Li/S) rechargeable battery (see Appendix D).
- Develop smart hybrid systems with batteries and energy conversion power sources. Key issues: predictive models of sink demands; duty cycles; air-breathing and air-independent operational modes.
- Develop generic modeling capabilities. Key issues: model materials properties through system integration; model steady-state and transient power source/sink behavior; establish control algorithms that optimize energy use.
- Develop BOP components for small fuel-cell systems. Key issues: reduce parasitic power, reduce size and weight, increase reliability, decrease costs.
- Develop small fuel processors for logistics fuel, methanol, ammonia, and other viable fuels. Key issues: thermal management; coking; sulfur removal; gas stream cleanup; start-up and load following; shutdown; packaging; interfaces with energy converter.
- Develop and field-test PEM/H_2 systems. Key issues: logistics impact of packaging hydrogen fuel; greater than 6 percent hydrogen storage/generation technologies; high-performance stacks.
- Develop and field-test DMFC hybrid systems. Key issues: logistics impact of packaging methanol fuel; low-crossover membranes; low catalyst loadings; high-activity catalysts; low-temperature storage and start-up.
- Conduct battlefield-relevant safety testing of alternative solutions (H_2, MeOH, ammonia, JP-8, and Li batteries).

There are two mid-term objectives:

- Develop rapid start-up, compact SOFC operating on low-sulfur logistics fuel or surrogates. Key issues: coke formation both inside and outside the stack; metal corrosion; integration with fuel processors; sulfur tolerance and/or removal.
- Develop complete small IC and Stirling engine systems with low signatures operating on JP-8 or diesel fuels. Key issues: bearings; BOP technologies—for example, fuel vaporization/atomization, control, acoustic and thermal signatures.

There are, as well, three far-term objectives:

- Develop high-specific-energy, air-breathing battery systems. Key issues: validate advanced battery concepts such as Li/air (TRL 3) and C/air (TRL 2); test and evaluate thermally self-sustaining C/air systems.
- Develop microelectromechanical system (MEMS) components for power technologies. Key issues: evaluate impact of incorporating MEMS in power systems; establish performance metrics and cost analysis for MEMS-based components; integrate MEMS components with conventionally fabricated components in complete systems.
- Develop SOFC systems that operate on high-sulfur fuels.

100-W Average Power

Total system mass for a 100-W average power mission is plotted against 24- and 72-hr mission lengths in Figures 2-4 and 2-5, respectively. The numerical data are given in Table 2-4. The battery technologies chosen for comparison are identical to those chosen for the 20-W case.

The DMFC (Giner, TRL 7) data are based on a 150-W packaged system that has not been optimized for low mass. In addition, the DMFC produces 10,800 Wh (150 W for 72 hr) at 17 percent efficiency. To produce the same number of watt-hours as the comparable technologies in the figure—that is, 7,200 Wh—would require 3.4 kg less fuel. A significantly lower mass for this system should be possible considering that a 100-W system would be lighter and that the efficiency of this system is not state of the art. A 30 percent total system efficiency has been demonstrated for DMFC by others (Ball, 20-W DMFC). The PEM/H_2 6 percent data are derived from the performance of a packaged, field-tested 100-W system (Ball, PPS100) and assume a 6 percent hydrogen storage/generation system. (Hydrogen stored at 5,000 psi with a safety factor of 2.25 will yield 6 percent hydrogen based on total mass.) As in the 20-W case, the IC data (projection based on scaling of the DSTAR 50-W engine/generator set, TRL 4) and the Stirling data (projection based on scaling of a Sunpower engine/generator set, TRL 1-2) are based on assumptions about the BOP items and scaling laws.

All of the energy conversion systems perform better than rechargeable batteries for the 24-hr mission at 100 W average power. Energy converters that employ JP-8 fuels (the SOFC and the IC and Stirling engines) are the best performers and are significantly less massive than even the best primary batteries in this regime.

For the 72-hr mission, energy conversion systems are an attractive alternative to batteries and could offer fivefold to tenfold mass reductions. As stated previously, there are significant operational constraints when using these air-breathing systems. From a systems perspective, the most mature technologies are PEM and DMFC, which require hydrogen (PEM) or methanol (DMFC). The advantage of the SOFC and the engine systems is their ability to operate on logistics fuels. SOFC offers the potential for the quietest system of the three operating on JP-8; however, some fuel processing is required before the fuel-cell stack. Small hydrocarbon fuel reformers under development (Altex) could be integrated with the SOFC, or integral CPOX reactors could be used. The latter have been demonstrated with butane (AMI). Small engines can utilize JP-8 type fuels directly; their minimal impact on logistics makes them very attractive. The reliability of these small engines remains to be determined.

Peak power in these systems could be provided by the prime power source, depending on the duty cycle. If the latter is too demanding, capacitors or lithium ion batteries could supplement the prime power source for the 200-W peak.

Findings for 100-W Average Power Regime

- Fueled systems become more attractive as power demand increases.
- Li ion batteries have the rate capability to power laser designators.
- Energy conversion technologies can reduce significantly the mass of 100-W systems that operate for 24 hours or longer.
- JP-8-fueled systems appear to weigh the least; however, these systems are not yet at high TRLs, and they may be more massive, have higher thermal and acoustic signatures, and be less reliable than other options.
- Non-JP-8-fueled systems offer significant performance advantages over batteries without some of the compromises of JP-8-fueled systems; however, the logistics burden will be greater to supply these fuels.

Science and Technology Objectives

S&T objectives consistent with the committee's selection of alternatives in the 100-W regime are listed below. The objectives are listed in priority order, along with key development issues to be resolved.

The three near-term objectives are these:

- Develop smart hybrid systems with fuel cells and high-power batteries or electrochemical capacitors.
- Develop small fuel processors for logistics fuels, methanol, ammonia, and other viable fuels. Key issues: thermal management, coking, sulfur removal, gas stream cleanup, start-up and load following, shutdown, packaging, and interfaces with energy converter.
- Evaluate DMFC and PEM systems for various specific missions. Key issues: modeling the capability of these systems with respect to loads, mission profiles, and operational and logistical constraints; overcoming technical issues as mentioned above.

There are two mid-term objectives:

- Develop small engines. Key issues: balance-of-plant (fuel delivery, vaporization, atomization, control system); integrating and packaging complete systems for field evaluation; validating performance scaling laws; assessing reliability and failure modes.
- Develop solid oxide fuel cells. Key issues: fuel processing, sulfur tolerance.

The sole far-term objective is this:

- Develop high-specific-energy, air-breathing batteries. Key issues: validating advanced battery concepts such as Li/air (TRL 3) and C/air (TRL 2); testing and evaluating thermally self-sustaining C/air systems; increasing the rate capability of Li/air by a factor between 5 and 10.

1- to 5-kW Average Power

System mass versus total energy is plotted in Figure 2-6 for three power source systems of TRL 9: 0.9 kW, 1.2 kW, and 2 kW. The plot was normalized to total energy delivered as these power sources have different power ratings. The conclusions below apply to power levels up to 5 kW. The Honda gasoline generator and the Mechron 2-kW diesel generator are commercial products (TRL 9). The latter is the generator supplied by the Project Manager for Mobile Electric Power (PMMEP) to the Army. The Ballard Nexa 1.2-kW PEM fuel cell is being packaged as a commercial product (TRL 9). A 70 percent efficient diesel fuel reformer and a 37 percent efficient fuel cell system were assumed in order to generate the Nexa fuel cell plot.

Mass was not estimated for the reformer, because no data were available. This reformer mass would need to be added to the points for the PEM fuel cell in order to obtain the mass for the total system. While the mass of the reformer is unknown, it would have to be 44 kg to result in a system having the same mass as the 2-kW diesel unit. It is anticipated that a reformer capable of supplying sufficient hydrogen to a 1-kW PEM fuel cell would have much less mass than this. The PEM fuel cell with an efficient diesel reformer could have an overall efficiency of 26 percent (0.7×0.37).

A system has not yet been demonstrated, so the Stirling engine was not included in Figure 2-6. Sunpower offers a 1-kW prototype Stirling engine having a dry mass of 32 kg,

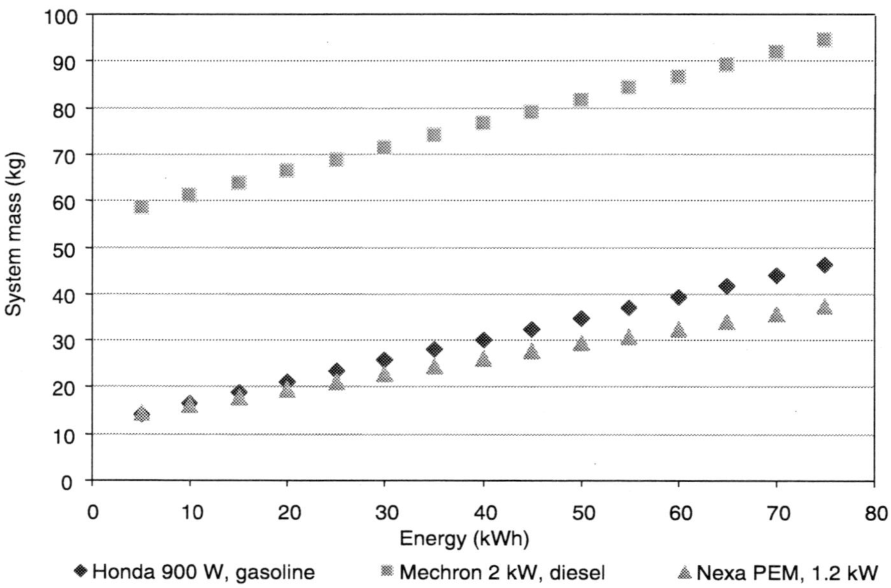

FIGURE 2-6 System mass versus total energy.

but this does not include the circulating cooling system or the burner and other ancillaries. An efficiency of 28 percent is claimed for thermal energy in to electricity out.

Stirling is inherently low in acoustic signature (less than 65 dBa at 1 meter) but has two sources of thermal signature. The motor-alternator must be kept at a rejection temperature on the order of 100°C, which is comparable to the temperature of operation of PEM fuel cells. The second source of thermal signature is the exhaust gas from the combustion process, which is likely to be hotter. The temperature is determined by the thermal recuperator and whatever mitigation scheme can be employed to further cool the exhaust gas.

Diesel and gasoline generators in this regime are highly developed (TRL 9). The Honda generator (19 percent efficient) has excellent performance and is light, but it operates on gasoline and has a significant acoustic signature (59 dBa at rated load). The Mechron 2-kW generator (16 percent efficient) in current use by the Army operates on diesel fuel and has a dry mass of 56 kg and an acoustic signature of <77 dBa at 7 meters. Specifications are readily available on the company Web sites.

Findings for the 1- to 5-kW Average Power Regime

- Fuel cells offer lower signatures than IC engines.
- Stirling engines offer potentially low thermal and acoustic signatures.

Science and Technology Objectives

S&T objectives consistent with the committee's selection of alternatives in the 1-5-kW regime are listed below:

- *Near-term objectives.* (1) Develop lightweight, efficient, 1- to 5-kW engines that operate on logistics fuel (key issues: tribology, reliability, integrating combustion sources with Stirling engines, and reducing system mass) and (2) develop lightweight logistics fuel reformers.
- *Mid-term objective.* Integrate logistics fuel reformers with lightweight PEM fuel cells.
- *Far-term objective.* Develop SOFCs. Integrate logistics fuel reformers with SOFCs.

3

Power System Design

This chapter describes how matching power source technologies (sources) with particular electronics applications (sinks) can affect the energy efficiency of systems. It then discusses the pivotal role of hybrid power/energy systems and provides insight into how to optimize the energy efficiency of an integrated source-sink system.

It is important to match energy requirements for each power-using device with the characteristics of the power sources. Figure 3-1 depicts the idealized characteristics of a battery, where voltage and efficiency throughout the discharge cycle and the charge capacity are not affected by the rate of discharge. Real batteries and other real power sources do not have such idealized characteristics. Voltage outputs and efficiencies of power supply systems are strong functions of each device's design details and the attributes of the duty cycle that must be serviced by the device. To optimize a system, both the dynamic characteristics of the power supply as well as attributes of the duty cycle must be understood.

Examples of such design considerations are highlighted in the following sections. These examples portray the need for detailed understandings of the dynamic characteristics of the power supply system and of the duty cycle. Without such details and a good analytical model to evaluate options, it is not possible to create an optimized system.

The various sources of power have different characteristics, as depicted for fuel cells and batteries in Figure 3-2. This figure shows the variation in efficiency with power output for a typical rechargeable battery and a direct methanol fuel cell. As is shown in the diagram the efficiency of the direct methanol fuel cell is somewhat above 30 percent at its rated power. As power output from the fuel cell is decreased, the efficiency drops continuously until it approaches zero at very small power output levels. This drop in efficiency is caused by the power requirements of balance-of-plant (BOP) components needed to support the fuel cell processes. This drop in efficiency makes the fuel cell a poor choice for systems where much of the energy is used at power levels lower than the rated level of the fuel cell. The battery, on the other hand, has a relatively high efficiency across the variation in power output from rated power to zero output. This high efficiency is related to the fact that energy is not needed to maintain the battery operation. Thus, batteries become the most energy-efficient source where the power outputs must swing over a large range of values.

DYNAMIC POWER

The traditional approach for estimating power supply capacity is to measure power demand in major operational states (standby, idle, peak), estimate the fraction of time to be spent in each of the states (the duty cycle), and sum over the resulting weighted averages. While this approach provides an estimate of system capacity, it ignores the dynamic behavior of the power demand. The power system can be more effectively designed if the dynamic profile of utilization is taken into account.

For example, a pager has very low power drain when in monitoring mode. Even when interfacing with a user to retrieve messages or to send a message the power requirement is easily met by a small battery. However, when the pager must transmit, the power required is substantially higher. Rather than construct a power supply to support the peak demand, pagers employ an electrochemical capacitor, which can be trickle-charged to provide large amounts of energy when the short message is transmitted. This hybrid power solution is smaller and weighs less than a system designed to meet the capacity of the transmission spikes.

Figure 3-3 depicts nonideal behavior, wherein voltage decreases with the time of discharge and the charge capacity decreases with discharge rate. In some battery chemistries, some capacity is recovered between periods of discharge.

It is common in small, portable electronic devices to reduce battery weight by drawing from the high demand portion of the battery output curve, causing a drop in

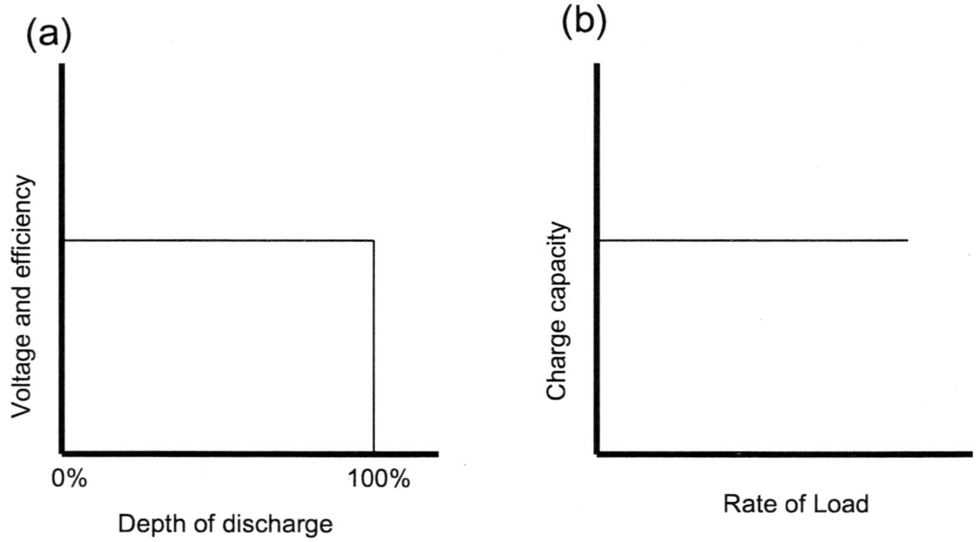

FIGURE 3-1 Characteristics of an ideal battery: (a) constant voltage and (b) constant capacity.

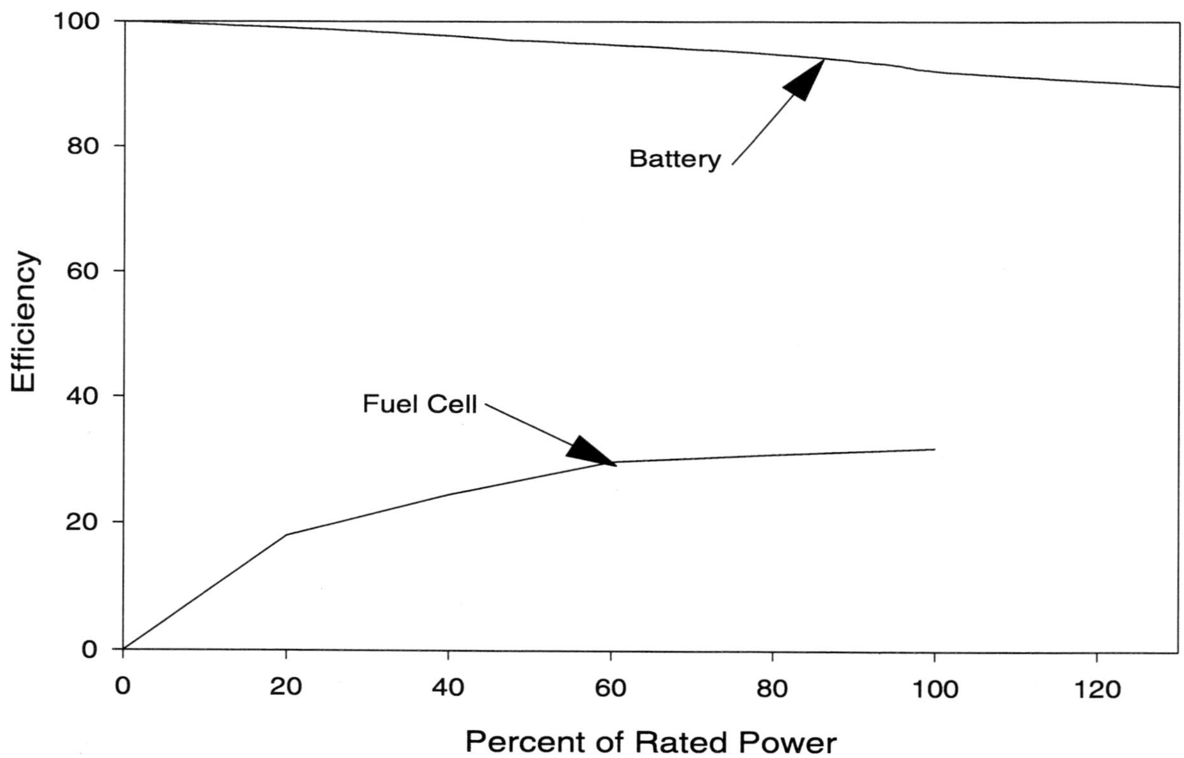

FIGURE 3-2 Power source efficiency variation with load. Typical efficiency for DMFC and battery. SOURCE: Adapted from data from Ball Aerospace 20-W fuel cell.

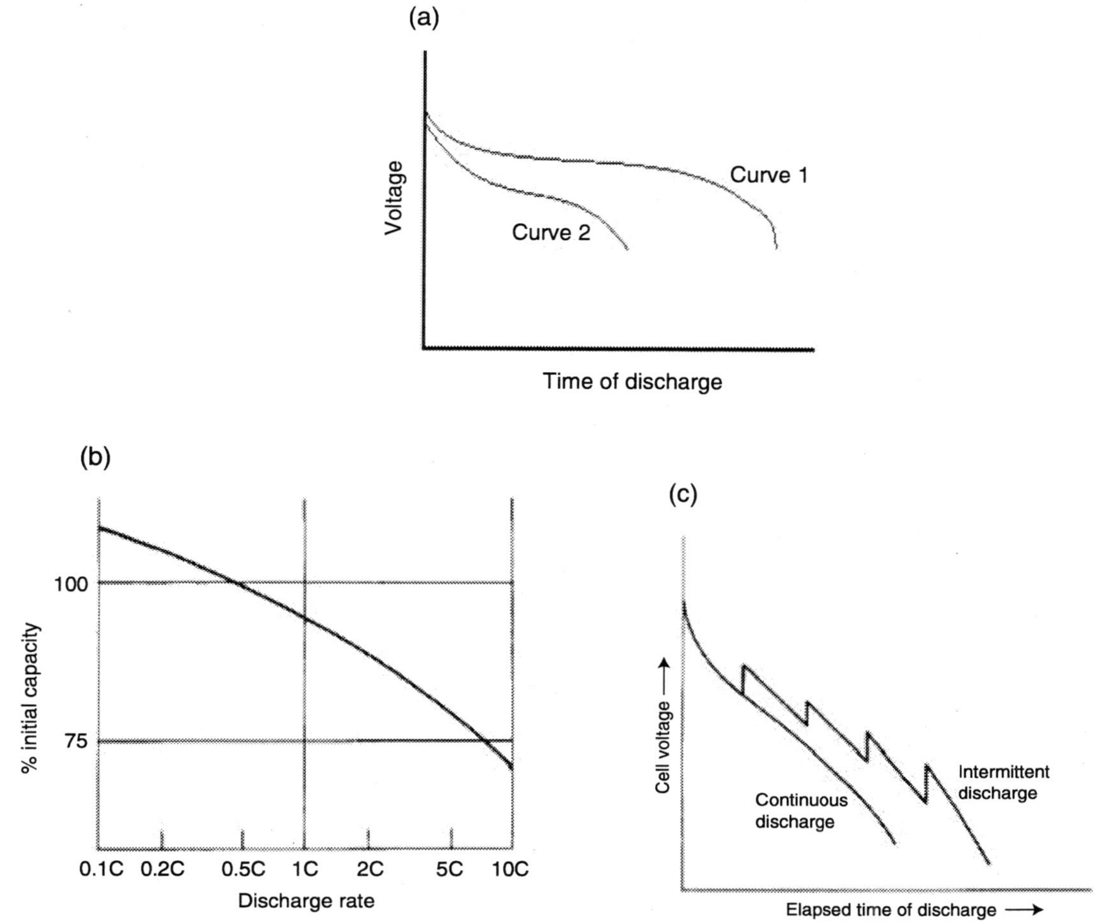

FIGURE 3-3 Typical voltage discharge profiles. Nonideal battery properties: (a) voltage change; (b) loss of capacity; and (c) recovery. SOURCE: Adapted from Linden, 1995.

capacity. Figure 3-3 shows that for a real battery, the rate of discharge affects the battery's apparent capacity. Figure 3-3a shows that voltage drops more rapidly for the higher discharge rate represented by curve 2 than for the lower discharge rate represented by curve 1. Figure 3-3b compares the percent of initial capacity and how it is affected by discharge rate. If C represents the capacity available at a standard discharge rate, then discharge rates from 0.1C to 10C are portrayed on the abscissa. As can be noted, initial capacity drops from 100 percent at 0.50C to approximately 70 percent at 10C. Figure 3-3c shows that if the discharge is intermittent, some recovery in cell output voltage occurs between periods of discharge.

Figure 3-4 presents another analysis of a battery system used on various duty cycles. For continuous discharge, the battery specific energy drops from approximately 138 Wh/kg at a specific power of 75 W/kg (point *B*) to about 90 Wh/kg at a specific power of 300 W/kg (point *A*). When intermittent operation is added, it is the peak power rather than the average power that determines capacity. In Figure 3-4, point *B* and point *C* are loads with the same average power, 75 W/kg, but the battery delivers a much higher capacity for point *B* than for point *C*. The reason for this is that the load of point *B* is a constant 75 W/kg, but the load of point *C* is an intermittent discharge with a peak power of 300 W/kg and a 25 percent duty cycle. The capacity delivered for the intermittent discharge of point *C* is much closer to the capacity delivered for a continuous discharge at the peak power value of 300 W/kg (point *A*) than it is to the capacity for a continuous discharge at the average power value (point *B*). Using the average power, the capacity would be estimated at 140 Wh/kg, nearly 30 percent greater than the actual capacity.

There is a slight dependence on the duty cycle, as shown by the family of curves representing intermittent loads with peaks of 100, 200, and 300 W and duty cycles of 25, 50, and 75 percent. However, despite this dependence, for intermit-

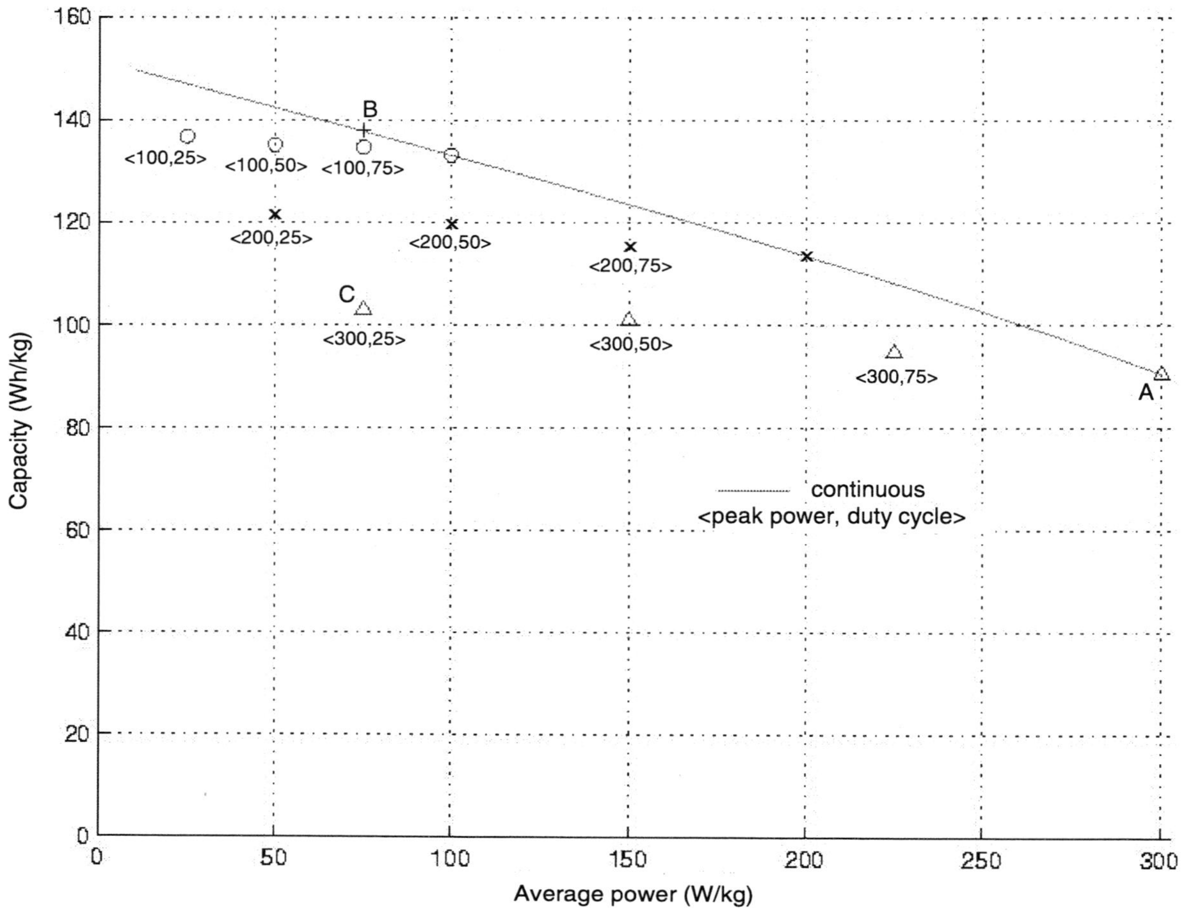

FIGURE 3-4 Doyle's Li ion model results for capacity versus average power, showing difference between continuous and intermittent loads of the same average value. SOURCE: Martin et al., 2003.

tent discharges, the capacity at a continuous discharge of the peak power is a better estimate than the capacity at a continuous discharge of the average power.

The typical power demand of mobile systems usually includes several periods of peak demand interspersed with potentially long periods of very low demand. Figure 3-5 depicts such a measured power demand of a speech recognition system on a mobile platform. Each peak represents processor and disk activity while recognizing a sentence.

HYBRID CONCEPTS

A hybrid power source usually combines a high-energy/low-power component with a low-energy/high-power component. Examples of hybrid power sources are battery + battery (e.g., Li ion + Zn/air), battery + capacitor (e.g., Li ion + electrochemical capacitor) and fuel cell + capacitor combinations. The rationale for hybrid power sources is to leverage the high-energy component with the high-power component, extending mission life and enhancing power capability while minimizing system weight or volume. Atwater et al. (2000) have demonstrated hybrid power sources with Li ion + Zn/air, Zn/air + electrochemical capacitor (EC), fuel cell + EC, and fuel cell + lead-acid combinations. In the Li ion + Zn/air hybrid, it was shown that the combined mission life (based on a communications equipment load profile) of the hybrid is almost six times longer than that of the individual components. In terms of specific energy, the hybrid had 198 Wh/kg, compared with 126 Wh/kg in the Li ion battery and 177 Wh/kg in the Zn/air battery.

Optimization of a hybrid power source is very complex, and the optimized combination of power sources might enhance only for a certain range of load regimes. Thus, it is highly desirable to develop a model that can predict and analyze the performance of various combinations of power

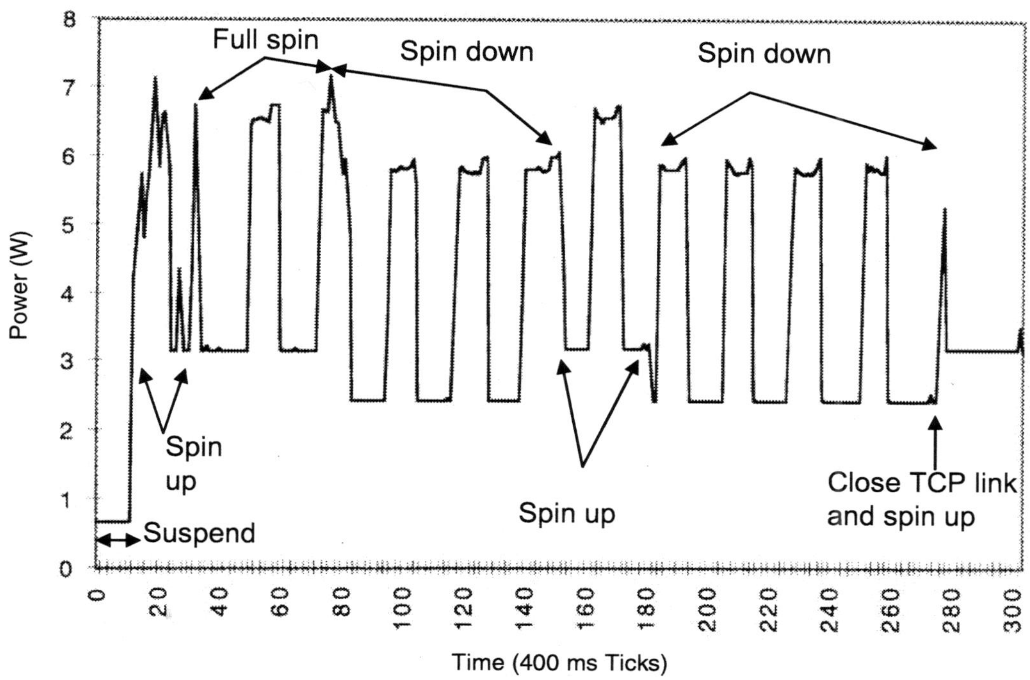

FIGURE 3-5 Power profile of a user interaction with a mobile computer. Each of the spikes represents the translation of a sentence from English to Serbo-Croatian, at which time the processor and disk are concurrently operating at full performance. NOTE: TCP, transmission control protocol. SOURCE: Reilly et al., 2000.

sources before one actually fabricates a hybrid. Further work should be carried out to develop models for various combinations such as battery + battery, fuel cell + battery, and fuel cell + capacitor hybrids.

In addition, effort should be devoted to understanding critical factors such as self-discharge and temperature effects. It is well known that capacitors generally have a higher self-discharge rate than batteries. Thus, for a battery + capacitor hybrid, the capacitor will drain energy from the battery, reducing mission life. Similarly, Zn/air cells do not perform well at low temperatures. Thus, for a hybrid power source involving Zn/air as one of the components, the low-temperature effect might necessitate a bigger Zn/air cell to achieve the same performance as at room temperature.

HYBRID ANALYSIS FOR THE SOLDIER SYSTEM

The component electronics of the soldier system will operate on duty cycles that vary significantly in power needs, so there are varying needs for high specific power and high specific energy. Appendix C of *Energy Efficient Technologies* (NRC, 1997) presented data on the effective use of a battery + capacitor hybrid system when high but short-duration power demands are required by a system. These characteristics are usually not found in the same power source. An exception is the battery + capacitor hybrid, which yields system weight gains and improved performance when peak power demand is of short duration (10 milliseconds or less). When the peak power demand is of longer duration, a hybrid might combine a fuel cell and a battery.

To highlight the benefits of such an approach, the committee analyzed the characteristics of a possible fuel cell + battery hybrid system using a direct methanol fuel cell (DMFC). While a system with any fuel could have been used for this analysis, a DMFC system was chosen because detailed full-load and part-load operating characteristics for such a system were readily available. A hydrogen fuel cell with comparable system-specific power, specific energy, and efficiency would also display benefits if used in such a hybrid combination.

The direct methanol fuel cell is a device with a high specific energy. This is a direct result of the fact that it is a fuel conversion device, and the energy stored in a typical fuel is at least an order of magnitude higher than the specific energy available in batteries of even the most advanced chemistry. On the other hand, the typical battery has a capability for higher specific power level than the typical fuel cell. For these reasons, hybrid systems with combinations of power sources are very attractive to satisfy overall power needs, especially for long missions.

Applicability of Hybrid Technologies

In the 1997 NRC report, several fueled hybrid energy systems were compared with current battery technology on the basis of source mass as a function of mission time in kilowatt-hours (NRC, 1997). In general, these calculations were done for roughly a 50-W system, and the most obvious conclusion to be drawn from the data is that there is a break point where it is more mass-efficient to use batteries than fueled systems. This is due to the dead weight of the converter, which is always present. Anytime the mass of batteries for a given mission is less than the dead weight of the converter, it is more advantageous to use a battery. As the kilowatt-hours for a given mission time increase, the dead weight becomes negligible compared with the fuel weight, and when that happens, if the converter is efficient, the fueled system is much less massive than a comparable battery. As the specific power of converters improves, the point at which converters are a more appropriate choice for a mission than batteries will move to shorter mission durations.

This simple analysis is not all inclusive, and other factors must be considered when comparing single-type energy systems with fueled hybrids. The most energy-dense single-type source is batteries based on lithium technologies where specific energy is approaching 200 Wh/kg. Hybrids offer enormous advantages from a simple energetics point of view for longer mission times. Conversion at a modest 20 percent of the lower heating value (LHV) of the fuel leads to specific energy factors 2 to 5 times better than those of the best primary batteries.

From a military standpoint, there are two important categories of hybrids: those that are air-breathing and those that are not air-breathing. In the latter category fall (1) the battery + battery hybrid (an extremely high specific energy and low specific power battery such as the lithium/$(CF)_x$ technology, coupled with a more conventional lithium battery having low specific energy but high specific power) and (2) the battery + electrochemical capacitor hybrid, successfully used in consumer electronics. The other category of hybrid (the air breathing) includes fueled and air-breathing combinations, perhaps coupling a rechargeable battery with a metal/air battery (such as zinc/air, aluminum/air, lithium/air), a motor generator (internal or external combustion driven), or a fuel cell (PEM/H_2, DMFC, SOFC). Table 3-1 is a compilation of the advantages and disadvantages of each of the technology types for attributes of interest to military applications.

Figure 3-6 analyzes the potential of such a hybrid system. The components chosen for an optimum system will depend heavily on the duty cycle demands of the soldier system combined and on the power system component characteristics. Commercial organizations that deal with these power management problems have simulation programs that use detailed system attributes for the potential power sources along with the details of the power sink demands and duty cycle of the system to be designed. These analytical programs are proprietary parts of their design group toolbox. Such high-fidelity simulations are key to their ability to introduce new products rapidly to the market.

To highlight the value of such analyses, Figure 3-6 depicts the system weight requirement for three candidate systems for meeting the 72-hr mission requirements for a prescribed duty cycle. Since no actual duty cycles are available for the soldier system as yet, the duty cycle chosen was for a 20-W average demand cycle with periodic peak demand of 50 W effective 10 percent of the time. The system chosen for analysis was an advanced direct methanol fuel cell combined with an advanced rechargeable battery. The battery was sized to meet the peak power demand of the duty cycle (50 W). The fuel cell was chosen to meet the mean power level required for the duty cycle. Both devices were among the candidates that could be available on an intermediate time horizon for applications to soldier systems.

Figure 3-6 shows that for this arbitrary 72-hr mission, the fuel cell + battery hybrid clearly outperforms either the battery or the fuel cell individually. The fuel cell weighed approximately 1.8 times as much as the hybrid system. The weight of the battery-only system was almost 2.5 times the weight of the hybrid system. The weights for the batteries were determined by using the target mission energy requirement. The limiting variable was that the required specific power had to be easily handled by the battery. For the fuel cell-only system, the dry system weight was assumed to be 2.5 times the 1.75 kg weight of a 20-W system, so that specific power was assumed to be similar for the 20-W and 50-W systems. Additional assumptions in the analysis were these: batteries were available in 0.5 kg sizes; fuel canisters for the fuel cell weighed 0.6 kg and supplied 800 Wh of energy for the full-load condition on fuel cell; for the 40 percent load condition in the fuel cell-only case, efficiency degrades, so that fuel canisters yield 612.5 Wh (Figure 3-2).

Although the analysis above was for an arbitrary duty cycle, the 2.5 ratio of peak to mean power and the 10 percent time at peak power seem reasonable for the demand of typical soldier applications. The weight benefit accrues as a function of these two ratios and becomes larger as the peak to mean power ratio increases and the duration of peak power shrinks as a percentage of total time. Thus, benefits would disappear as peak to mean power approaches 1 or as the percentage of time spent at peak power approaches 100.

Looking at Figure 3-6, it is apparent that for short missions the power source of choice based on system weight would always be a battery. As battery specific energy values improve, the length of mission for which they outperform other system alternatives will increase.

Soldier systems operate in a unique environment, characterized by the extremes of dismounted warfare. They must be "ruggedized" to withstand physical punishment and not pose an extra hazard for the soldier under enemy fire; they must operate in various climes, including conditions of sand and dust; they must have simple controls, enabling such

TABLE 3-1 Comparison of Single Battery versus Hybrids for Attributes of Importance in Military Applications

Factor	Single Type: Battery	Hybrid: Non-Air-Breathing (Battery + battery hybrids)	Hybrid: Air-Breathing (Fueled System + Metal/Air Battery)
Effects of environment	Minimal. Typically operates from -30°C to +70°C. (However, low temperatures significantly reduce specific energy of battery unless it is warmed with use.) Disposal problems. Orientation-independent. Can be submerged.	Typically operates from -30°C to +70°C. (However, low temperatures significantly reduce specific energy of battery unless it is warmed with use.) Disposal problems. Orientation-independent. Can be submerged.	Needs preheating to operate at low temperatures. Some units are orientation-dependent. Minimal disposal problems. Performance altitude-dependent (but has been shown to work where humans work at >15,000 ft). Sensitive to dust and pollutants in air. Special precautions required for liquid immersion.
Logistics	Some restrictions on transport of lithium technologies. Many suppliers offshore. Readily available for civil applications.	Some restrictions on transport of lithium technologies. Many suppliers offshore. Readily available for civil applications.	Logistics dependent on fuel. Systems operating on logistics fuel are immature and at low technology readiness levels. Logistics not in place for other fuels such as H_2, methanol, natural gas, aluminum, zinc, and carbon.
Logistics infrastructure	Logistics infrastructure in place to deal with lithium technologies.	Logistics infrastructure in place to deal with lithium technologies.	Logistics infrastructure would have to be developed to implement. If widely accepted, energy-efficient systems will reduce logistics burden.
Versatility/utility	Extremely versatile. Many sizes possible. Can be adapted to power almost anything. Enormous range of sizes and shapes. Has limited specific energy with little room for improvement.	Must always have two units. Able to provide higher power to limit of power-dense unit. Duty cycle determines relative sizes of two units. Operates as a battery trickle charger.	Operates primarily in a battery charger mode, but can provide power directly to load up to rated limit. Ultrarapid recharging of primary energy store. Must shut off all inlets and outlets if immersed. Some versions highly sensitive to dust and pollutants in air. Special procedure for low-temperature operation. Acoustic, thermal, and chemical signature problems. Three to eight times more energy dense for long missions.
Safety	Safe at low specific energy and low discharge rates. At high specific energy and high discharge rates, units may explode or rupture, dispersing toxic chemicals.	Safe at low specific energy and low discharge rates. At high specific energy and high discharge rates, units may explode or rupture, dispersing toxic chemicals.	Fuel-dependent safety. Reactants are separate. Some units are hot, presenting fire hazard if fuel spills. Not inherently explosive.
Reliability	Highly reliable. High specific energy/power units have safeguards to prevent explosive rupture events.	Highly reliable. High specific energy/power units have safeguards to prevent explosive rupture events.	Motor generator sets in civil and military applications have excellent ratings in larger sizes. Insufficient data to estimate reliability for small sizes and for various fuel cell systems.
Manufacturability	Large civil infrastructure currently manufactures batteries in an enormous range of sizes and shapes.	Large civil infrastructure currently manufactures batteries in an enormous range of sizes and shapes. At present the electronic and software infrastructure to operate optimized hybrid systems does not reside in the military sector.	Small motors and generators have established manufacturing infrastructure to produce large quantities. Little market for small fuel cells; hence manufacturing infrastructure is limited. Market demands will establish infrastructure.
Availability	Readily available in commercial sector. Military needs not always met (as for *all* energy technologies that are to be used for military applications). Special tooling and facilities may be needed; these come at a premium price. Having materials infrastructure helps reduce some costs of an integrated power system. But ultimately, packager and systems integrator must provide government with a military-class system based on those materials.	Readily available in commercial sector. Military needs not always met (as for *all* energy technologies that are to be used for military applications). Special tooling and facilities may be needed; these come at a premium price. Having materials infrastructure helps reduce some costs of an integrated power system. But ultimately packager and systems integrator must provide government with a military class system based on those materials.	Small motors and generators up to 1 to 2 hp are widely available at low cost. Stirling is emerging, but established market and manufacturing infrastructure exists for cooling applications. Fuel cells in this size range are one-off, with no market incentive to develop mass market manufacturing capability.

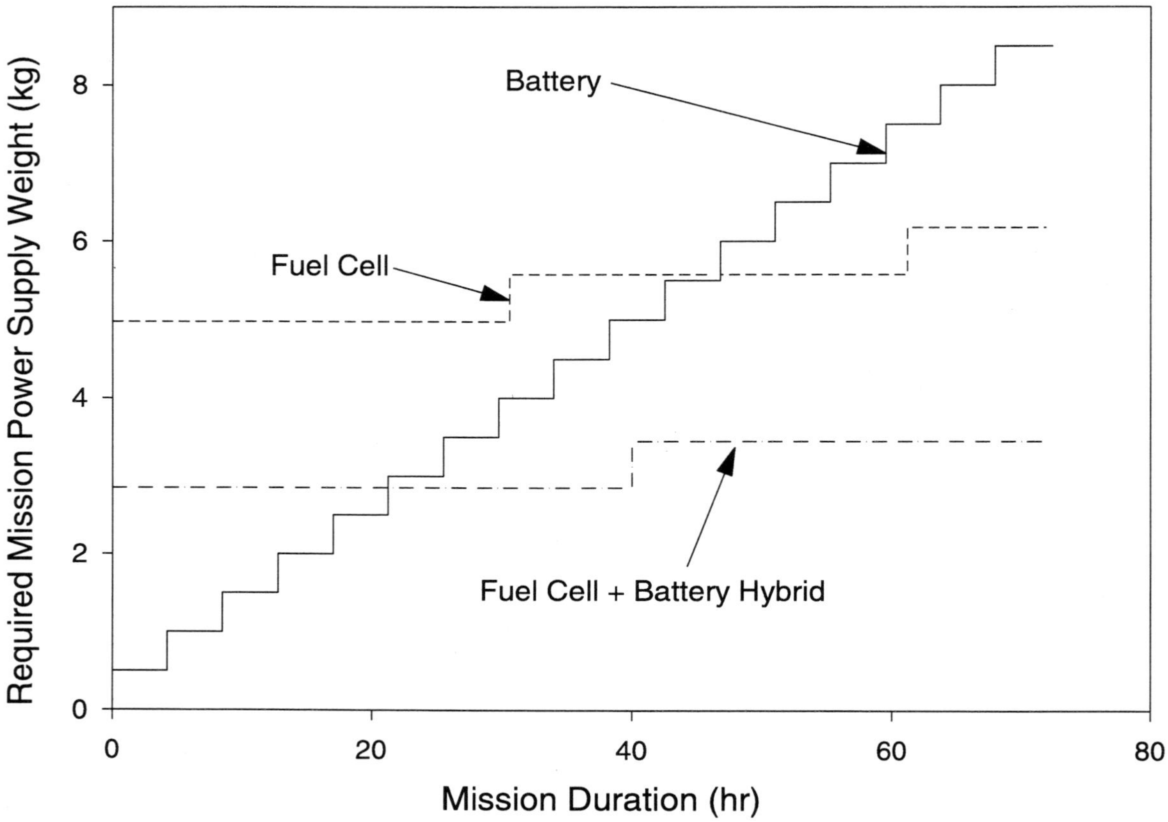

FIGURE 3-6 Soldier power demand for 20-W average, 50-W peak 10 percent of the time. Performance comparison for batteries alone, fuel cell alone, and hybrid battery + fuel cell. Data are based on Ball Aerospace 20-W DMFC and Li ion rechargeable battery and assume fuel cell system dry weight changes directly with peak power required and fuel packaging is the same for both large and small fuel cell.

things as a stealth mode to evade detection; and they must be waterproof.

For this reason, a key consideration in developing a hybrid system is the amount of time submerged with no access to air. Hybrids must be smart in that they can automatically close any air ports to protect against intrusion. These requirements would necessitate the inclusion of a battery or other non-air-breathing source to supply power when submerged operation is required.

Battery + Battery Hybrid

The committee was provided data on an experimental demonstration of the characteristics of a Zn/air + Li ion battery combination proposed for use in the Objective Force Warrior-Alternative Technology Demonstration (OFW-ATD), as shown in Figure 3-7 (Graham and Feldman, 2003). This comparison is flawed in that it does not consider the weight attributes of such a combined system. In general, battery + battery hybrids show an advantage over a single battery system only if the energy battery is incapable of meeting the power peaks required by the mission. This can be evaluated by calculating the specific power required of the energy battery to produce a given peak power, then dividing it by the weight of the energy battery, sized according to the total energy required for the mission.

In an example provided by the Army for OFW, the estimated peak power required was 40 W (Graham and Feldman, 2003). The energy source was a Zn/air battery having a specific energy of 300 Wh/kg. A 24-hour mission would require 14.82-W average power for 24 hours, or 356 Wh, corresponding to a Zn/air battery weighing 1.095 kg. The peak specific power at which the Zn/air battery would be required to operate would be 40 W/1.095 kg, or 36.5 W/kg, which is well within the capabilities of a Zn/air battery. Thus, there would be no need for a high-specific-power battery for this particular mission. If the peak specific power required is much higher, then a high-specific-power battery may help. In this example, a more useful comparison would be the discharge time for a Zn/air battery of the same weight as the hybrid system. Here there would be no significant difference in discharge times, though the voltage dip during high power pulses would be smaller for the hybrid.

Table 3-1 illustrates the differences between single and hybrid sources for several performance categories. In addition, effort should be devoted to understanding critical factors such as self-discharge and temperature effects. It is well known that capacitors generally have higher self-discharge rates than batteries. Thus, for a battery + capacitor hybrid, the capacitor will drain energy from the battery, shortening mission life. Similarly, Zn/air cells do not perform well at low temperatures. Thus, for a hybrid power source involving Zn/air as one of the components, low temperatures might necessitate a larger Zn/air cell in order to achieve the same performance as at room temperature.

SYSTEM CONFIGURATION CHOICES

The choice of system elements must consider the specific characteristics of both the energy source and energy sink elements of the system. The wide range of energy sinks includes standard computer and display hardware using powers of milliwatts to watts; laser target designators demanding 100 W or more; soldier cooling hardware demanding tens of watts continuously; and exoskeletal devices that demand very large amounts of power and energy.

The proposed sources of energy include primary batteries (with high energy densities and modest internal ohmic resistance); rechargeable batteries (with lower energy density and lower ohmic resistance); fuel cells (with low energy output but high energy density related to their use of high-energy fuels); and engine-driven generators (with high outputs and high energy densities but problematic noise and heat signatures).

Hybrid systems have the ability to improve system power usage by limiting the voltage drops that are imposed on the power supplies for higher current demand duty cycles. This ability reduces the heat generated by ohmic resistance in the power supply. Additionally, such hybrid combinations can avoid the transitions into low effectiveness operation of one of the power supply components. See Figure 3-7 for a comparison of data for the Zn/air battery curve as one example.

These power sources might be used at different times or in combination to best satisfy soldier needs. To determine the most appropriate combination for satisfying a soldier's

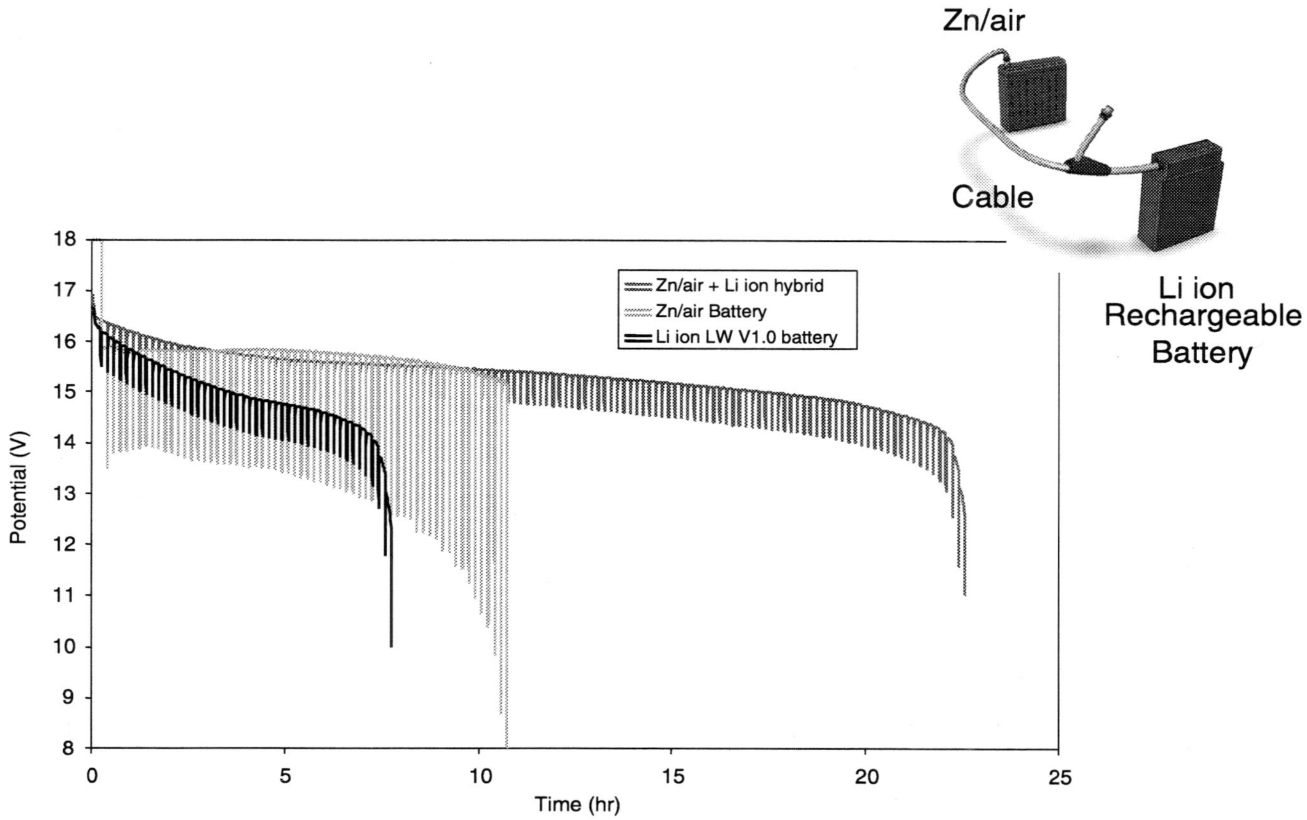

FIGURE 3-7 Performance of hybrid as compared with performance of single components in power load cyclic profile of 9 min, 12 W, and 1 min, 40 W. Hybrid's gain is approximately 4 hours. SOURCE: Graham and Feldman, 2003.

mission needs, one must consider source and sink characteristics along with mission requirements and duty cycles.

Any two sources may be combined into a hybrid to satisfy soldier system needs. As described previously, the combination of rechargeable batteries and fuel cells might be used to meet periodic high current demand and high energy needs in combination. The combination of an air-breathing generator and rechargeable batteries might be better for a mission that requires a soldier to be immersed in water.

Matching Source with Sinks

The Army has defined several mission scenarios, and each must be validated so that appropriate systems can be selected confidently. Before any system is selected, it is important that the combination of energy source, energy sink, and soldier mission requirements and duty cycles be considered jointly.

Comparing power source performance metrics under identical load conditions and operational scenarios allows for the best assessment of energy alternatives. This requires knowledge of the power demand and time data for every piece of equipment for every soldier for a statistically significant number of missions. The OFW-ATD Program includes a modeling effort that predicts power demand by positing mission scenarios, estimating duty cycles, and using power sink specification data for all components of the Land Warrior soldier system. In addition, the OFW Program plans to monitor the power demand of specific components during actual or simulated missions. The information gathered will be used to validate the models and provide realistic boundary conditions for total energy, average power, peak power, and duty cycles for various missions. The validated models should lead to more effective planning and designs. For example, the optimal suite of energy storage and energy conversion devices, fuel quantities, etc., could be determined for each mission.

MODELING REQUIREMENTS

Commercial developers of power systems simulate the power use of their systems so they can rapidly optimize hybrid system choices. These modeling efforts require detailed descriptions of the power supply characteristics of the candidate components and are considered to be key proprietary parts of their in-house design processes. The dynamic response characteristics of components—that is of the available battery types, capacitors, and fuel cells—as a function of current flow rate and ambient conditions must be known to obtain accurate results.

The Army will need a similar modeling approach in order to make appropriate system design choices. The type of duty cycle encountered in real field applications is key to acquiring an optimized design and must be determined through experimentation. The Army should invest in such a modeling capability, which would be essential for effective power management at the system power input level. This capability could be combined with other power management capabilities focused on system power output and power demands.

Researchers at the University of South Carolina have developed modeling software known as the Virtual Test Bed (VTB), with the goal of optimizing the usage of charge storage devices for specialized applications. They can input the parameters for general battery systems and then study how the battery will perform under specified loads. Software can also be used to model hybrid power sources, where a battery is used for low power and an electrochemical capacitor is used for pulsed power. With this approach, the South Carolina team successfully improved power utilization for a device that utilizes a hybrid system (Dougal et al., 2002). The results are largely nonintuitive, and extensive modeling was needed to identify the optimum power source.

It is clear to the committee that high-fidelity modeling will be needed to optimize the Army soldier system. For models to be useful, the power and energy usage of a mission must be specified. However, until the OFW system has been put in place and usage data gathered, the power and energy inputs to these models are not available. The mission requirements are especially critical when pulsed power is needed, in which case a battery + capacitor system might be useful; also, the duty cycle of the pulsed power must be known. For instance, a minute-long pulse can be delivered efficiently with a battery, but a capacitor + battery hybrid power source might be better for a device that senses at low power and then transmits data using a millisecond pulse. Because engineers of military equipment are often looking to adapt new technology, they resist modeling efforts until they have their system design completed. Although the modeling effort can be time consuming, it should not be delayed.

One challenge with the modeling approach is to develop code with general rules that can be rapidly adapted by systems engineers to help guide their choices toward components that may save power in the overall system. Also, the models need to be able to take into account the behavior of real systems—that is, systems that fade in performance over time or have a range of performance values.

In summary, modeling has the potential to save time and money in the development of efficient portable electronic systems if accurate system inputs can be supplied. The modeling can complement experimental data as it narrows down the parameters of optimization. Any power solution ultimately needs to be verified with experimental data, but modeling can expedite selection of the power source. Ideally, the military should develop and acquire new equipment based on recommendations and considerations gained from power sources modeling, so that the lifetime of the equipment can be maximized.

4

Soldier Energy Sinks

This chapter describes power demand characteristics of the electronics needed for soldier applications—the soldier power sinks. It discusses both low-power and high-power electronics applications (the latter include laser designators, microclimate cooling, and exoskeletons) in the regimes specified in the task statement. The chapter concludes with observations on the impact of Army logistics on the selection of power solutions for the soldier.

As has been discussed, the energy-consuming hardware used in soldier applications can be roughly divided into two categories: the silicon circuits for computation and communications and the highly specialized sensors and transducers for the weapon systems. The total power demand of the soldier system is now more or less equally distributed between these two categories if the duty cycle is taken into account.

LOW-POWER ELECTRONICS TECHNOLOGY

Communications and computation circuitry both make use of silicon integrated circuit technology, which is being driven by market requirements to be ever more energy efficient. But there are major differences in the degree to which communications and computing can exploit commercial advances to reduce power requirements. The challenge for military communications and computation circuitry applications is how best to capitalize on commercial advances.

The exponential growth that has characterized improvements in commercial processor performance may dramatically slow over the next 10 years. (See the section "Commercial Trends" in Chapter 5.) In fact, the energy efficiency of the best-performing processors has begun to decline as they move into a power-limited operating regime. This is so because the impossibility of removing the heat that is created by the processor at the highest possible clock rates and voltages by means of heat sinks makes it impossible to exploit the peak performance of these processors. While it is believed that the soldier requirements are well below this limit, it is important to be aware of the limitations of the technology, which will become more severe with time.

There is a hierarchy of computation architectures for each computation application. For signal processing (video compression, data communications) the energy efficiency can vary over many orders of magnitude. In order of energy efficiency, the least efficient are standard microprocessor architectures (even with power optimizations such as X-scale). Specialized processors that can implement signal processing functions can be up to 10 times more efficient. Field-programmable gate arrays are even more energy efficient since they can implement highly parallel solutions. If full flexibility is not required, then application-specific integrated circuit (ASIC) solutions can be employed; these provide another order of magnitude improvement in efficiency.

HIGH-POWER APPLICATIONS

Specialized applications such as weapons, infrared sights, laser designators, the exoskeleton, and microclimate cooling require individual consideration. These applications in most cases depend on nonsilicon technology and are highly specialized for their function. Case-by-case investigation of the systems used for each application is needed to identify the devices that consume the most energy and therefore present the best opportunities for power reduction.

Laser Designators

Laser designators represent a unique power demand that can be many times greater than that of other electronics for the Objective Force Warrior (OFW). Systems now available to special operations forces operate at system voltages of 18 to 30, with a standby mode that draws 10 W and an active mode requiring 180 W. The active mode could last as long as 10 to 40 seconds, and specifications indicate that the battery complement required is five BA 5590/U lithium or four BB 590 NiCd batteries. Because these requirements are far in excess of the other OFW requirements, the energy efficiency of such systems should be of special interest to the Army. Power requirements for new laser designating devices could easily override other considerations and control the selection of a centralized energy source for the soldier system.

The Army did not provide its concept for fielding the devices, and the power demands were not included in the power allocation for the OFW. Clearly, reducing power demand should be a major consideration in designing laser designators. The committee expects that the energy efficiency of such devices can be improved, but it did not know enough about such devices to recommend specific improvement techniques.

Microclimate Cooling Systems

Although soldier microclimate management efforts have been under way since the early 1990s, they remain in an early stage of development. Specifications, including duty cycles and energy use data, for the many prospective systems being contemplated by the Army vary considerably, and the committee was unable to determine whether the solutions proposed are particularly energy efficient. In this report, the committee does document its observations on the preliminary design information provided by the Army program manager (Masadi, 2003). A variety of system approaches—from ice cooling systems to vapor compression and absorption refrigeration—have been tried.

The basic difficulty with management of the dismounted soldier's microclimate is the large power requirement for such an effort. A dismounted soldier doing very light work such as guard duty has a work rate between 100 and 175 W. Light work such as cleaning a rifle has a work rate of 125 to 325 W. Moderate work such as foxhole digging has a work rate between 325 and 500 W. Heavy work such as emplacement digging has a work rate above 500 W. Since the human body is on the order of 18 percent efficient, these work rates would require cooling rates five times greater.

The power required for microclimate cooling is much greater than the power required for other functions of the dismounted soldier. Because of these excessive power requirements, microclimate cooling for the OFW will be limited to providing ventilation for soldiers clad in protective clothing, whereby a ventilator moves air into and through the soldier's protective clothing to provide modest improvements in comfort.

This effort is being accomplished in concert with modifications to soldier uniforms that provide passages near the skin in which air can flow to provide evaporation and transport of perspiration. This air movement can be provided with minimum pressure drops for filtering toxic or noxious inputs to the system or modest pressure drops to afford airflow within the soldier's uniform. For the OFW program, a power budget of 10-12 W has been allocated to accomplish ventilation and cooling tasks. It is expected that a separate specific power source will be provided for the ventilation system.

The present approach is to think of providing soldier microclimate cooling in three variations. The first variation, described in the two preceding paragraphs, is passive cooling designed into a porous uniform. The second variation would provide active cooling by ventilating the soldier's uniform with continuous airflow from a small fan. The third variation would provide active cooling with a mechanical refrigeration device. The second and third variations are not likely to be available until significant development has been completed. This means that active cooling will not be available until well after the OFW system is completed.

Researchers at the University of Wisconsin surveyed a broad range of possible alternatives for use in active cooling of the dismounted soldier. They also looked at potential power sources for the alternative systems. Their analysis used performance characteristics of large, state-of-the-art, optimized cooling system components, even though such components do not exist on the scale required for individual soldier use. When such components are scaled, there might be significant degradation of system performance.

Vapor compression systems for cooling, which might be made available soonest, were among the most effective. Other systems that might have comparable effectiveness were in early development and had correspondingly low TRLs. The vapor compression systems were projected to be capable of a coefficient of performance of 3.8 and would thus consume approximately 80 W on a continuous basis to handle a modest 300 W body heat load.

Findings

Forced ventilation systems are the most energy-efficient cooling system, but their capabilities are limited. This type of microclimate cooling is only effective under favorable conditions of temperature and relative humidity. Because of the effectiveness of ventilation systems and their modest power requirements (10 to 12 W) they are being developed before active microclimate cooling systems. Even this relatively low power level may require a separate power source, in addition to those sources envisioned for the other soldier electronics.

If active microclimate cooling is to be pursued, much higher power levels will be required. Based on the power demands for microclimate cooling options explored in the University of Wisconsin survey, an energy-dense fueled system will probably be required, even for advanced systems.

Exoskeleton Systems

The Defense Advanced Research Projects Agency (DARPA) is developing an exoskeleton prototype as part of a human performance augmentation program that is focused on developing load-carrying devices that will increase the speed, endurance, and load-carrying capacity of soldiers in combat environments. This program, like the microclimate cooling efforts, is in its early development phase, and the committee did not analyze the relative merits of solutions on the drawing boards. The committee does, however, document here its observations on preliminary approaches presented to it by DARPA (Main, 2003). Specific target applications include moving heavy loads over rough terrain, bearing heavy weapons or equipment, carrying and powering breaching equipment, and using the exoskeleton as a platform for increased body armor. The vision is to utilize such devices with power supplies that are energetically autonomous of other power sources. Such exoskeleton devices must mimic human motions and provide close human/machine integration. Human/machine interfaces must provide for transparent control of the exoskeleton over extended periods of operation.

Since the loads carried by rifle-squad personnel during a 72-hr mission are projected to range from 140 pounds for a rifleman to 185 pounds for an antitank specialist, such human augmentation is clearly desirable. Developing augmentation devices that are both compatible with and comparable to human capabilities is a daunting task. Human muscles provide large motions and high repetition rates. Various power sources are compared in Figure 4-1 on the basis of the stress/strain product capabilities versus their frequency response capabilities.

As can be seen in Figure 4-1, human muscle provides a combination of high-frequency and high stress/strain activity.

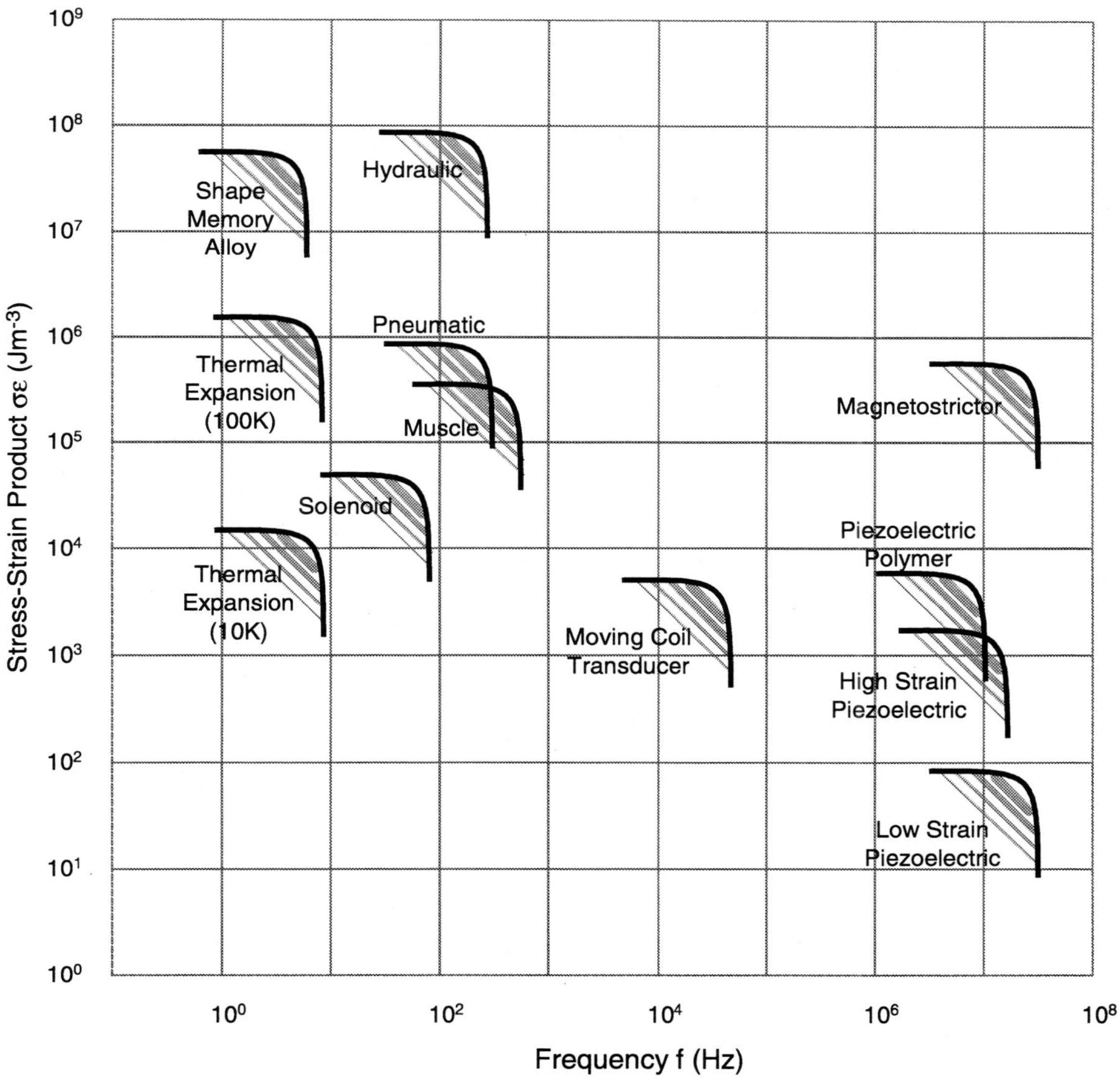

FIGURE 4-1 Comparison of various means of exoskeletal actuation on the basis of stress/strain product capabilities. SOURCE: Main, 2003.

If an exoskeleton device is to improve on human performance, it must match the range of speed capabilities and provide load-carrying capabilities in excess of human muscle capabilities. Figure 4-1 compares various means of exoskeleton actuation and indicates that hydraulic actuation systems provide the best approach. The actuation capabilities must then be combined with a power source of appropriate specific power and energy.

In 1994, the Prospector VI workshop (AUSRI, 1994) considered most forms of electrically driven actuation and concluded that hybrid power and energy sources, generally a high specific battery combined with an electrochemical capacitor, offset many of the problems of high-maintenance hydraulic systems. The workshop did not, however, consider specific exoskeleton applications where bandwidth and stress-strain product were figures of merit.

Choices for the power supply to such exoskeleton devices are limited to those of extremely high specific power and specific energy. The generalized Ragone plot in Figure 4-2 show that combustion engines are the most viable candidates. Thus, a combination of an engine-driven hydraulic compressor and hydraulic actuators could provide the energy levels and the frequency response required for exoskeleton operation. The DARPA program manager indicated that the size, weight, and inertial characteristics of electrical actuators that would be needed to provide equivalent energy levels for these devices would be too great. If one reviews the capabilities of electrohydraulic and electromechanical

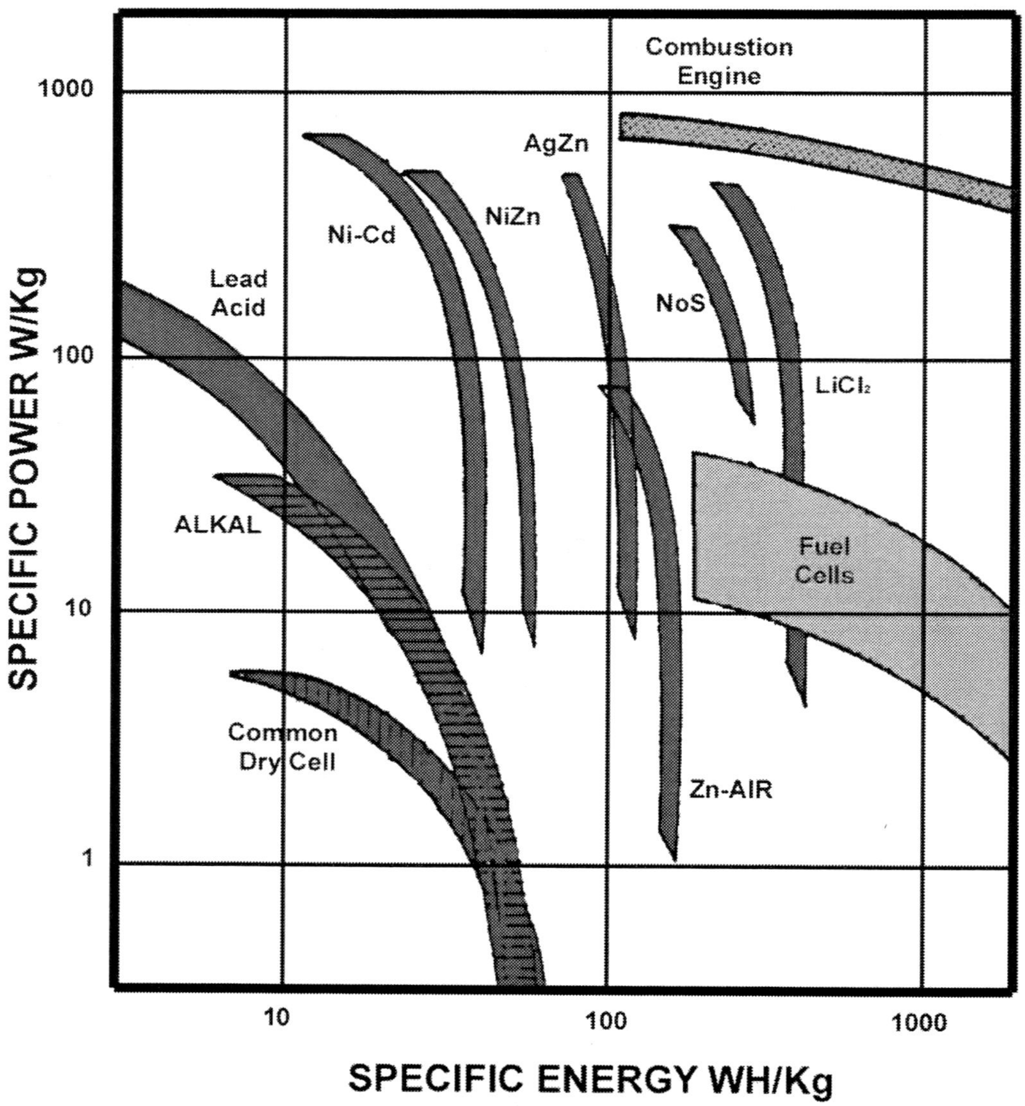

FIGURE 4-2 Generalized Ragone plot of different power sources. Only engines have both the high specific energy and high specific power needed for exoskeletal devices. SOURCE: Main, 2003.

actuators in a recent NASA review of actuation systems, one draws a similar conclusion (Merryman and Hall, 1996).

Problems remain with the application of hydraulic power. If one sizes a hydraulic system for maximum output in terms of power and frequency response, large energy losses may occur in the control valves when the system is operating at low energy output. In electrical systems such problems are handled with pulse-width modulation of actuators. To minimize losses in energy and thus excessive heat loads in hydraulic systems operating over a wide range of loads and speeds, it would be necessary to modulate system pressures as a function of system output requirements. At present, controls and devices for such flow modulation appear not to be available. It is unclear whether such an actuation system could be created along with a load-following energy source.

Findings

To significantly enhance individual soldier capabilities, exoskeleton systems will require power systems with output in excess of 150 W. Such systems will require fuel-converting power sources with the power density and size of the best small engine power sources available today. These power sources must be integrated into efficient hydraulic actuation and control systems.

The need to function over a range from high to low will require the development of load-following systems and control technologies not available today. The efficiencies of present hydraulic systems peak at the highest output rates and fall drastically as work output falls, yielding low system efficiencies. To make these exoskeleton approaches viable, some load-following approach must be developed to overcome these problems.

BATTLEFIELD LOGISTICS

Choices for power solutions must be compatible with the Army's logistical and operational systems. In fact, it is unlikely that this study would have been commissioned were it not for the Army's concern about the logistics of soldier power. This section focuses on what is needed beyond hardware to provide power on the battlefield. It considers such things as the weight carried by dismounted soldiers engaging in multiple operational scenarios; how to transport the batteries, energy-producing fuels, or other consumables to the soldiers when needed; and the costs of acquiring and supporting the energy sources appropriate to different power demands.

Standardization

Enhanced electronics equipment and weaponry make the dismounted infantry soldier a formidable fighter, but the logisticians who support the soldier must now add electric power to the other essentials—ammunition, food, and water—that must be supplied to the battlefield. Complicating the issue, they must provide and support a growing variety of power sources.

In this, standardization is a logistician's friend. While the military standard for form, fit, and function is the BA5590 lithium battery, the most-used battery is a commercial AA cell. This stems from the obvious fact that the commercial battery can be found around the globe, even when the supply system falls short; they are so ubiquitous that they can be bought in stores on the economy or mailed from home.

Standardization extends to tactics, techniques, and procedures, as soldiers from the Ranger Regiment, Light Infantry divisions, Airborne and Air Assault divisions, new Stryker brigades, and Future Combat Systems will all be expected to fight dismounted from their diverse platforms. Fully training these individuals in all the complex skills needed in modern warfare becomes increasingly challenging. Therefore, while systems providing power and energy to the fully dismounted soldier must emphasize energy efficiency, the electronics equipment cannot be radically different from that employed by mounted soldiers in the armored, mechanized, Stryker, and other brigades.

It is much more difficult to deal with a multiplicity of parts or systems performing similar functions. Individual soldiers and units are less likely to run short of needed supplies when these supplies are fewer in type and larger in quantity. This means that minimizing the variety of batteries and other power sources should remain an important objective. Similarly, efforts should be made to minimize the need for specialized fuels.

The cost of providing disposable batteries has already grown burdensome and can only grow more so. Beyond the direct cost of batteries procured for combat is the even larger amount required for training. Training and fighting with a rechargeable power source, such as a rechargeable battery or a fuel cell hybrid, could reduce costs, but the Army prefers to train as it will fight and fight as it has trained.

By experience, the Army's operational ideal power source would be a single battery, lightweight and disposable, not harmful to the environment, able to withstand the rigors of combat without endangering soldiers, and capable of providing power for all soldier electronics. This fact must be kept in mind when considering alternative power sources.

Operational Considerations

The Lead Technology Integrator (LTI) for the Objective Force Warrior-Advanced Technology Demonstration (OFW-ATD) Program is integrating its prototype to support six operational scenarios (Erb, 2003). Modifying any of them would require justification and approval from the Army Training and Doctrine Command (TRADOC). While developers must have scenarios against which to be measured, the diversity of missions given to dismounted soldiers virtually assures that large standard deviations will exist. Simulations of operations, along with real-world measurements, may narrow these modeling deviations, but there will always be a broad spectrum of operations to be considered.

Military tactics, techniques, and procedures influence soldier power requirements on the battlefield. For example, soldiers do not have identical power needs. There are also organizational and hierarchical limitations on the use and placement of recharging facilities as well as limitations on taking full advantage of the most energy-efficient technologies.

The Army provides its soldiers with equipment that has been tested to assure reliability, and it builds redundancy into the supply system. Unfortunately, the contingencies of warfare often reduce planning time and can result in a "come as you are" fight. At this point, the electronics used and the loads necessary to support them will be determined by the equipment on hand. This makes open architectures and versatile power sources desirable.

In general, the electronics equipment and the power sources for the dismounted soldier should be engineered together as a system capable of performing with high reliability and human-factored to minimize interference with all conceivable postures and positioning. Equipment must emphasize small and light construction, preferably modular (i.e., capable of being included or discarded without adverse impact on other components of the soldier system). Not least, by design it should provide for graceful degradation, so that soldiers are not faced with all-or-nothing situations on the battlefield.

Mission Duration

The Army wants its dismounted soldiers to be capable of operating at least 24 hr without having to replace any power sources. It also wants the means to transport and sustain power sources for 72-hr operations.

An important major lesson that has been learned in Operation Iraqi Freedom is that just-in-time logistics resupply may not be able to keep up with the rapid tempo of military operations. Reliable delivery of consumables to every small unit every 24 hr was not achieved and appears unlikely to be achieved during this or similar engagements in the future.

The initial OFW concept envisions a rechargeable power solution, and soldiers will have to be trained to recharge in combat. To be effective, soldier power solutions must both reduce the logistics burden and be adaptable to combat operations.

Dismounted Soldier Load

At every level of operation the Army seeks to plan as far ahead as possible. This is true at all levels—fire team, squad, platoon, and company. Capable leaders in disciplined units will tailor the loads that their soldiers carry to best perform the anticipated mission; it is these leaders who ultimately decide how much weight is enough. If mobility and endurance are expected to be high priorities, loads will be lighter. In more static situations, they might be heavier.

In scenarios describing future operations, power sources account for a relatively modest percentage of soldiers' loads: Weapons, ammunition, water, protective armor, and special-purpose gear form the lion's share of the weight of the gear. Nevertheless, it is the responsibility of all who provide equipment to the dismounted soldier to make each item as small, light, and user-friendly as possible.

The committee was amazed to learn about the physical loads—at times in excess of 100 pounds—that were actually borne by dismounted soldiers in Afghanistan (Erb, 2003). These loads did not include weight that would be added by the LW ensemble.

Army concepts for the future LW assume that LW effectiveness and survivability will depend on situational awareness, communications, and special-purpose equipment and on weapons that depend heavily, and in some instances almost totally, on reliable power. For example, missions lasting 24 hr will require the soldier to carry a certain weight of primary batteries to provide this power. Barring resupply, longer 72-hr missions would require at least three times as much battery weight; this additional weight is heavy enough to subtract substantially from the amount of mission equipment that can be carried.

There will be times when missions last longer than anticipated or when resupply is delayed, and it will then be necessary for the soldier to make do. Two important capabilities are needed and should be part of any power solution: reliable indicators for energy remaining in the power sources and provision for graceful degradation, which will enable a soldier to minimize all but the most essential power consumption needed to complete the mission.

Perhaps the clearest indication that graceful degradation is needed is the desire expressed by soldiers that individual weapons have a separate power source rather than being tethered, with or without wires, to the LW ensemble. Regardless of what else happens, soldiers want to be able to fight and to defend themselves.

Recharging on the Battlefield

Dismounted units in any active state will need either a new supply of disposable batteries or a recharge of rechargeable ones frequently—OFW scenarios suggest as often as daily. However the recharge solution adds complexity. Instead of delivering fresh batteries and departing, soldiers manning forward supply vehicles will have to collect discharged units and provide for their renewal, either on board the vehicle or at a location further back. Recharging at the small unit's location virtually dictates the presence for a time of a charger-carrying vehicle, possibly a small robotic vehicle. While locating such battery rechargers on vehicles or in forward locations is a worthy approach, one must recognize that they could then become choke points that diminish a small unit's operational flexibility.

User/Developer Interface

Open and thorough interaction between user and developer is important in defining and developing any new Army system. This is especially so for a recharging system that will necessitate changes in training and doctrine. TRADOC, which represents the Army's combat units, legitimately wants the best that can be provided and will ask for it. However, user/development interaction is especially important in soldier systems, where the weight is so critical and where seemingly small changes in requirements can cause large increases in weight.

Acquisition Planning

The Department of Defense (DOD) acquisition system is so cumbersome that it is difficult if not impossible to match the rapid development of electronics and devices in the commercial sector. Slow development cycles, often coupled with the need for backward compatibility with earlier systems, drive the Army toward evolutionary rather than revolutionary approaches to development. This militates against special-purpose development and qualification, because by the time a revolutionary product is completed, it is one or more generations out of date.

Army dismounted infantry and special operations forces (SOF) have power/energy requirements (environmental, duty cycle, power levels) that cannot always be satisfied by equipment in commercial use. The Army and DOD then have no choice but to develop their own sources or adapt commercial ones. The cost of doing so becomes a serious factor.

Solutions appropriate for SOF, with their smaller structure and unique ability to provide logistics for targeted missions, are not necessarily appropriate for all dismounted soldiers. The number of non-SOF dismounted infantry is substantially larger, and they require more standardized items.

Energy-efficient electronics and reduced consumption of energy are less important to mounted (armored and mechanized infantry) soldiers, who are less dependent on portable electronics. There is little incentive to reduce power demand if better performance can be achieved using non-portable electronics without concern for weight. For example, it is likely that radios developed for dismounted soldiers and SOF will continue to have quite different designs and require separate development and testing.

Industrial Base Issues

Maintaining suppliers, particularly suppliers of special-purpose power sources that have no commercial counterparts, will require effort. For example, small fuel-cell systems will continue to be viewed as special-purpose items until a design concept resolving all fuel issues is agreed upon. Even then, the relatively small quantities needed by dismounted soldiers limit the Army's procurement leverage and will continue to affect the amount of attention that can be paid to energy efficiency in adapted commercial designs. Fortunately, there is increasing interest in the commercial world for energy efficiency in small devices, and the Army must take advantage of commercial-off-the-shelf (COTS) opportunities that develop.

Energy efficiency was not an important factor in developing the original prototype for the LW ensemble. The OFW-ATD now appears to recognize its importance and must follow through with its choices of electronics. In acquiring electronics for the dismounted soldier, the Army needs to recognize that appropriate incentives will be necessary to encourage the development and use of low-power technologies and designs.

OFW Operational Assumptions

The Army apparently has made a rational decision to use OFW-ATD to provide near-term operational capabilities that can help to overcome past LW shortcomings. The program is fast-paced and resources are focused on quick iterations and results. While this is commendable, it means that OFW is not currently structured to advance the development of LW-unique technologies. One aspect of this that affects the power source solution is the emphasis placed on 24-hr vs. 72-hr mission lengths. The assumption that resupply will occur at least once a day may not be valid. The 72-hr mission requirement should drive the power source solution, but it is not being emphasized. Either the scope of the OFW-ATD Program must be increased or a separate effort must focus on the lengthier mission requirement.

Near-Term Considerations

Many considerations are involved in choosing between primary and rechargeable batteries for operational use. Primaries are attractive. As described in Appendix D, they have higher energy densities than their rechargeable counterparts. They are disposable after use and thereby lighten the soldier's load. Their principal drawback is affordability. Rechargeables are now used in training, because the cost of using disposables increased significantly as battery usage grew. The Army recognized that it could not sustain this peacetime expense.

Rechargeable batteries offer an attractive near-term solution. Their life-cycle costs are far more affordable, the weight and volume that the logistics system must bring to forward positions are reduced substantially, and the risk is low. Moreover, the supply line is radically shortened; no longer is there a need to reach back to the continental United States for battery resupply. A fully recharged battery is as close as the nearest recharger, which may be on the soldier himself, on a vehicle in the squad, or within his unit.

The additional jet propellant 8 (JP-8) required to fuel rechargers is minuscule compared to amounts already in-theatre for vehicles and other engines. However, a decision to rely on rechargeable batteries also brings disadvantages and burdens. Less energy can be stored in a rechargeable than in a primary battery of equal weight. Rechargers in different configurations will introduce additional items of equipment that must be proliferated and supported in forward areas. These, as well as the batteries being recharged, will necessitate additional vehicles and possibly additional personnel. Vehicles such as the high-mobility, multipurpose wheeled vehicle (HMMWV, or Humvee) presently in use have preexisting priority weight and volume claims, and their alternators are ill-suited for supporting repeated recharging requirements. Unmanned robotic ground vehicles, which are scheduled to arrive on the scene, may provide at least part of the solution.

A seemingly attractive near-term solution to further reduce risk is to stockpile primary batteries for use only in emergency and combat situations, using rechargeable batteries for training. Field training exercises usually last 3 days or less, and recharging during most training could be handled administratively. This solution avoids the requirement for a fully developed recharger support structure for both training and combat operations.

However, another risk is introduced. Experience during the early weeks of Operation Iraqi Freedom, when battery stockpiles were almost exhausted, does not inspire optimism that adequate stocks of primary batteries can be procured and rotated during peacetime. If the Army believes it important to have this flexibility, it must develop soldier systems that can safely and easily use primaries and rechargeables interchangeably.

In the near term, there appears to be little risk that two fully functional rechargeable batteries cannot provide adequate power for the 24-hr missions foreseen today. Absent an ability to resupply or recharge during the mission, there is no technology solution that permits accomplishing a 72-hr mission without tripling this load of batteries to six, thereby adding significantly to the soldier's load.

Findings

Batteries have become critical elements of a soldier's combat effectiveness, so the Army must make distribution and recharger configuration choices that incorporate redundancy to protect against battle damage and overcome periods of nonavailability, as well as insure against choke points during periods of intense use. A reserve of fully charged batteries will always be a safety valve, with the size of the reserve dependent on the situation.

Rechargeable batteries are a feasible near-term solution for short missions but involve risks that must be mitigated. They should be regarded only as an interim step in an evolving progression. Where possible, the Army should ensure the inclusion of hardware "enablers" in the near term that will facilitate solutions in the long term.

Long-Term Considerations

A hybrid power solution that packages a fuel cell and a rechargeable battery in the soldier's load offers attractive increases in performance. Figure 3-6 showed that a full 72-hr mission requiring 20 W of average power and 50 W of peak power over 10 percent of the time can be satisfied with a lithium ion rechargeable battery and a direct methanol fuel cell (DMFC). One hundred soldiers operating under these conditions for 72 hours would require approximately 200 liters of methanol (120 kg plus packaging) to replenish their individual fuel cells. Consumption would amount to about two thirds of a liter per soldier per day. For shorter missions, it would be possible and desirable for soldiers to leave their fuel cell behind and rely on embedded and central batteries. With this solution, the need for separate battery rechargers would be eliminated, bringing logistical and operational benefits.

Absent is the ability for the Army to devote substantial resources to maturing the DMFC; it must depend on progress in the commercial market. Fortunately, more than one corporation appears close to offering these devices in the 20-W range for use in portable computers, and the opportunity to exploit COTS may soon arise.

Proceeding with the direct methanol fuel cells used in this example is yet another undertaking that demonstrates the dynamic tension between new technologies with increased capabilities and the logistics systems that must support soldiers in their use of the new capabilities in combat. Technologists are dissuaded from pursuing a promising new opportunity when they have little confidence that it will be incorporated into operational use. Logisticians have little reason to do comprehensive, detailed analyses of the logistics implications, advantages, and burdens of a new technology until they have reason to believe that it will be coming into the force.

While this hybrid approach has the potential to meaningfully reduce the cost, weight, and volume of the logistics chain, its burdens include adding another separate fuel, methanol, to the diesel and JP-8 found in forward combat units. In the near future, the fielding implications of fuel cell hybrid solutions should be examined to include operational benefits, life-cycle costs, and safety and risk.

Commercial investments and applications may make small fuel cells a more acceptable alternative for batteries.

Prototypes have already been demonstrated for use in laptop computers. The Army should be prepared to take advantage of such investments by evaluating the logistical trade-offs involved in fielding nonbattery power solutions. The evaluation should include timelines for the introduction of fueled system alternatives and estimate the consequent reductions (if any) in battery use. This would permit decision makers to make informed judgments on how operational advantages compare to the added burden.

Predictive modeling must play an important part in mitigating risk through analysis of this opportunity. For example, the miniscenario described above was overly simple, taking no account of duty cycles. With predictive models in place, multiple scenarios can be reviewed. Predictive models will be invaluable in examining alternatives, understanding the impacts of unique operations, and narrowing the standard deviations in complex scenarios.

Tethering a single power source to multiple components, such as weapons, sensors, laser designators, and helmets, presents unattractive human factors issues that must be resolved. On the other hand, the attractiveness of a one fuel cell/one battery solution cannot be denied. One way to reap the benefits of both untethered subsystems and a master central power source is to colocate the batteries with the components that require relatively large amounts of power and energy. These could then be tethered to the central source with quick breakaway connectors to permit recharging from the fuel cell during periods of relative inactivity and unhampered action when the need is greatest.

Findings

The one fuel cell/one battery central power approach with satellite rechargeable batteries appears to have great promise. If a detailed analysis justifies its operational value and logistics supportability, it is a candidate for accelerated development.

Hybrids have great potential, both as power sources and, over the longer term, as factors in decisions on centralized vs. distributed power. The packaging of a nonstandard fuel for the fuel cell requires an immediate and thorough analysis of trade-offs. The fuel cell approach can then be pursued in earnest if this screen is passed. To facilitate the analysis, the Army should use predictive modeling to narrow sigmas and evaluate choices.

Efficient fuel cells permitting use of JP-8 or reformed fuels could offer even greater advantages and more relief from the logistics burden, eliminating the need for packaged nonstandard fuels from the supply chain. Combining an air-breathing fuel cell with a rechargeable battery would enable operation in all conditions, but would have to be integrated into a much smaller package than current state-of-the-art to be viable on the battlefield.

5

Progress

This chapter reviews the progress that has been made by the Army over the past 5 years since the publication of *Energy-Efficient Technologies* (NRC, 1997). It discusses reasonable expectations for the near- and far-term Land Warrior (LW) ensemble based on improvements observed from prototype through Stryker Interoperable (SI) to the Objective Force Warrior-Advanced Technology Demonstration (OFW-ATD). The chapter also reviews and extends predictions of the previous report and describes significant changes in commercial development trends.

OBJECTIVE FORCE WARRIOR-ADVANCED TECHNOLOGY DEMONSTRATION

Overall the Board on Army Science and Technology (BAST) committee was impressed by the amount of effort that has been put into power and energy concerns for the Objective Force Warrior-Advanced Technology Demonstration (OFW-ATD) Program. The Army, through its contracted lead technology integrator (LTI), now has a process that allows developing prototype technology to be inserted into the LW program. Initially, the focus is on relatively near-term technologies that can be demonstrated in the OFW-ATD and later inserted into a new version of Land Warrior, Advanced Capability (LW-AC). A snapshot of the concepts being considered for the next LW by the LTI showed the committee the direction of Army power solutions and soldier requirements.

The OFW design team is focusing on a 24-hr autonomous mission. It is assumed that soldiers will be resupplied at least once every day and that each unit will have the means (possibly a robotic vehicle) to carry extra batteries, rechargers, or fuel needed for 72-hr missions. These assumptions led to a goal for the future soldier fighting load as low as 50 pounds, including no more than 2 pounds for a power system providing 12 W.

The weapon for the OFW will be cabled for recharging through the centralized source and detached when fighting. A hybrid power concept is being considered for the power source to support extended missions. It involves a high specific energy/low specific power system for steady loads and a high specific power/low specific energy system for peak demands.

Comparison of OFW Concepts with Land Warrior

Peak power demand anticipated for the LW ensemble will not change appreciably under the initial OFW-ATD concept. Table 5-1 also shows that estimates for LW-SI peak power did not change substantially from the original LW estimates in *Energy-Efficient Technologies* (NRC, 1997). Similarly, average power estimates for the ensemble have remained relatively constant. As shown in Table 5-2, the peak power, average power, and average/peak ratios (without radios) for three generations of LW (LW, LW-SI, and OFW) are within 20 percent of one another.

The committee's analysis was organized around four categories of functions that make up the suite of LW electronics, including displays, computer subsystems, sensors, and communications.

Displays

There has been a substantial reduction in the power requirement of the helmet-mounted display from the 1997 LW to the proposed 2007 OFW. The other displays consume approximately the same power in spite of power reduction progress made in reducing power needed for displays. This is probably the result of added capabilities (e.g., color and higher resolution) anticipated for the recent LW versions.

TABLE 5-1 Comparison of Estimated Power Requirements of Land Warrior System, by Function (All Peak Power)

Function	Land Warrior, 1997[a] (W)	Land Warrior (Stryker), 2004[b] (W)	Objective Force Warrior, 2007[c] (W)
Communications			
Soldier radio	7.4	5.97	6.2
Squad radio	14	7.8	7.8
UAW/robotic vehicle			6
Computer displays			
Handheld flat panel	6.4	7.04	7.05
Helmet-mounted	4.9	1.4	0.5
Integrated sight—module display	2.6	2.65	3
Sensors	7.9	16.75	9.5
Computer	14.8	15.7	17.42
Total	58	57.31	57.97

[a]Estimates from NRC (1997).

[b]Breakdown of Stryker interoperable wattages (Brower, 2003)

- Soldier radio, 2.5 (WLAN card) + 0.17 (WLAN digital radio) + 3.3 (WLAN antenna and amplifier);
- Squad radio, 7.8 (leader radio);
- Handheld flat panel, 6.29 (display) + 0.75 (keyboard);
- Helmet-mounted display, 1.4;
- Integrated sight display, 2.15 (thermal weapon sight) + 0.5 (daylight video sight);
- Sensors (everything but the 25-W chemical agent detector), 4 (multifunction laser) + 0.2 (weapon user input device) +1.17 (card reader) + 0.17 (GPS interface) + 2.4 (GPS card) + 0.25 (dead reckoning) + 0.5 (microphone) + 0.76 (helmet integrated assembly) + 0.6 (laser detector) + 2.5 (slave hub processor) + 0.76 (computer USB hub) + 0.32 (slave hub) + 2.97 (weapon hub) + 0.15 (chemical agent detector); and
- Computer, 12 (computer processing card) + 3 (DRAM and radio frequency conversion) + 0.7 (computer/master hub subsystem).

[c]Breakdown of OFW wattages (Erb, 2003):

- Soldier radio, 7.8 (JTRS numbers not available; assumed the same as Stryker MBITR radio);
- Squad radio, 4.4 (communications processor card) + 0.6 (WLAN card) + 0.6 (VoIP processor) + 0.6 (WLAN antenna);
- UAW/robotic vehicle, 3 to 10 W for como-crypto interface (Brower, 2003);
- Handheld flat panel, 6.3 + 0.75 (handheld KB and cable);
- Helmet-mounted display, 0.5;
- Integrated sight, 3 (HIA, module including breakaway connection to body PAN);
- Sensors, 2.15 (thermal weapons sight) + 1.1 (daylight video sight) + 4 (multifunction laser) + 1.5 (GPS) + 0.25 (dead reckoning module) + 0.5 (microphone/speaker assembly); and
- Computer, 2.1 (computer assembly) + 10.9 (computer processing card) + 3.42 (body PAN hub) + 1 (PAN weapon hub).

NOTE: DRAM, dynamic random access memory; GPS, global positioning system; HIA, high integration actuator; JTRS, Joint Tactical Radio System; KB, kilobyte; MBITR, multiband intra/inter team radio; OFW, Objective Force Warrior; PAN, primary area network; UAW, universal access workstation; WLAN, wireless local area network; USB, universal serial bus; VoIP, Voice over Internet Protocol.

TABLE 5-2 Comparison of Estimated Peak and Average Power and Their Ratios for Land Warrior Systems

	System (Without Radios)		
	Land Warrior, 1997[a]	Land Warrior (Stryker), 2004[b]	Objective Force Warrior, 2007[c]
Peak power (W)	35.3	43.59	37.97
Average power (W)	15.35	19.5	15.8
Average/peak ratio	0.435	0.447	0.366

[a]Estimates from NRC, 1997.

[b]Estimates from Brower, 2003.

[c]Estimates from Erb, 2003.

Computers

The computer subsystems evolved in a different way. As opposed to the single processor in the earliest LW system, the OFW design includes a number of processors that are interconnected through multiple high-speed local area networks (body LAN). To assure longevity, the OFW design has gone to an open architecture with several standardized buses (e.g., Firewire, gigabit Ethernet). While a variety of buses enhances the number of modules that could connect to the OFW electronics, it does exact a premium for power to keep all the buses energized, even if there is only one transaction type per bus. The result has been that computer power demands of the three generations of LW are about constant in spite of the significant improvement in energy efficiency of the underlying computer system technologies.

Sensors

Power demand estimates for the sensor suite have increased slightly over the three generations. The number and types of sensors are similar, but there have been significant improvements in functionality.

Communications

The OFW-ATD is working to develop power-aware applications and an intelligent middleware layer that will efficiently manage bandwidth usage. It will use a radio (Joint Tactical Radio System (JTRS) Cluster 5 SLICE radio) that creates a peer-to-peer network architecture, but the software-based design solution for JTRS may not allow for reductions in power demand. Initial OFW estimates take an optimistic approach to what will be available in 2007 by using power numbers no worse than those for the MBITR radio with LW-SI, thus enabling a complete high-level comparison of the overall power demands of LW electronics. The importance of soldier communications-electronics to reductions in power demand is discussed further below.

LAND WARRIOR POWER IMPROVEMENTS

The Army Program Executive Office (PEO-Soldier) and the LTI provided briefings on facets of the OFW-ATD relating to power, including power sink technologies, system design, doctrine, networking, and power sources. For each recommendation derived from the five conclusions in *Energy-Efficient Technologies* (NRC, 1997), the LTI provided specific examples of how the OFW-ATD would improve the energy effectiveness of the LW system.

Application of *Energy Efficient Technologies* to the OFW-ATD Program

Efficient power usage is understood to be critical to the success of the OFW-ATD and has been identified as a key performance metric. Every energy-consuming capability must earn its way onto the system. The OFW-ATD is using state-of-the-art technology developed by both commercial entities (for main computers) and government programs (for radios) to reduce system power demand. The OFW-ATD system architecture is intended to be flexible enough to incorporate new technology as it is developed.

The OFW-ATD is also using computer-aided design, simulation, and profiling tools to perform power analysis on proposed designs. All applications to be developed will be power-aware. A custom Linux kernel is being created, tuned for power and security.

To synchronize doctrine with technology and minimize soldier communications transmissions, the OFW-ATD software team is working closely with the operational effectiveness team to analyze and prioritize the data that need to be transmitted across the Army force structure. The OFW-ATD is also studying the operational utility of unmanned vehicles, both air and ground, as nodes in the peer-to-peer network architecture that would have more powerful reach-back capabilities.

In response to the 1997 recommendation for research and development in rechargeable batteries, the OFW-ATD is working with a commercial battery supplier to devise a rechargeable battery with a specific energy on the order of 200 Wh/kg. The OFW-ATD is also tracking advances in energy sources and is exploring hybrid systems. It also supports continued research into advanced energy sources and is particularly interested in direct methanol fuel cells.

Committee Observations on Initial OFW-ATD Concepts

The assumption that future soldiers will be resupplied every 24 hr does not necessarily modify the goal of 72-hr self-sufficiency. To justify the 24-hr assumption, OFW makes the further assumption that each unit will use a vehicle for resupply (such as a robotic Mule). This assumption, endorsed by the Army, is important because it has substantial logistical implications, for including procurement and transportation of Mules, spare parts, and fuel.

The A123Systems battery technology is high payoff/high risk. The high power capability of this approach is based on doped $LiFePO_4$ as extrapolated from laboratory experiments. The chemistry is inherently safe, and the raw material is not expensive. Nevertheless, there are other rechargeable alternatives, and the OFW-ATD will probably need to pursue these concurrently to ensure success.

OFW plans to embed data-logging capability in the system to track energy and power demand. This is an excellent idea that could provide an initial basis for subsequent development of needed models. Actual power usage profiles can also be used to evaluate future design trade-offs. This should provide substantially more accuracy for power modeling than the current estimates of subsystem duty cycles.

The committee notes that hybrid techniques, such as

charging an electrochemical capacitor from a battery, can be used to provide pulses of energy for low-duty-cycle (1 percent or less) devices without compromising battery capacity. However, if devices like the multifunction laser (estimated at 4 W peak power) are used several times in succession, the local duty cycle will either not allow enough time for the hybrid system to recover or else appear as a higher duty cycle peak demand with adverse affects on capacity, especially if batteries are the main power source.

The committee had additional substantive observations on the length of the LW procurement cycle, incentives for saving power, and the important area of soldier communications.

Length of the Procurement Cycle

The time horizon of the LTI contract is probably not long enough to collect effective feedback even though the program has adopted an iterative development and improvement cycle. Further, the relatively short duration of the cycle militates against there being time to bring energy-efficient system-on-a-chip (SoC) technology into play and significantly reduce power demand of one or more of the LW subsystems.

Although SoC technology targeting the OFW application was recommended in *Energy-Efficient Technologies* (NRC, 1997), OFW-ATD will use off-the-shelf electronics componentry. The Army should begin the development of SoC technology that can evolve with requirements as they are understood and with new developments in algorithms and protocols.

The current approach to designing and procuring Army soldier systems should be contrasted with the approach to designing commercial products such as cell phones. Each generation of an Army system starts with a new contractor and a clean sheet of paper, allowing only an after-the-fact, lessons-learned critique of the previous generation. There is not a lot of learning transferred from one generation to the next, leading to a lack of continuity in design concepts. In the commercial world, by contrast, there is continuity between products over multiple generations. Commercial electronics developers aim for progressive improvements to design, with successive generations of SoCs containing capabilities better than those of the previous generation. By building on earlier SoC designs, the cost and risk of the later generations are substantially less than the cost and risk of the first generation.

Another cost of the standards-based plug-and-play strategy of the OFW is that standardized USB and Ethernet hubs for the wired soldier body LAN use considerably more energy and do not directly enhance the effectiveness of the soldier. There should be investments in developing low-power interconnect technology. For example, a fundamentally different approach would use a high-speed, short-range, wireless body LAN that has an undetectable emission signature. One approach to this goal would use ultrawideband (UWB) radio transmission, which is being developed commercially to transmit in the 3 to 5 GHz range and (from the 802.11.3a proposed standards) at a rate of 100 Mbits per second at less than 100 mW total power demand; receive at 200 mW; and, most important, have sleep modes that are three orders of magnitude lower (Batra et al., 2003). The very low transmit power and secure characteristics of the transmission signal would provide a radio frequency (RF) signature that would not be detectable beyond 10 m (or even less if the transmit power is constrained further). Chip sets for this network approach are projected to be available from Motorola, Intel, and other vendors for home video networks in the next 3 to 4 years at a cost of less than $5 per node. Assuming that internal or local operational interference is not a problem, such wireless technology could obviate the tethering of data buses on a soldier's system.

Incentives for Reducing Power

A lesson from the original LW integration program is that there was not enough time or money to fully optimize energy efficiency. Due consideration must be given not only to the various power sources and sinks, but also to designs for electronics integration and power management.

The OFW LTI will propose systems for integration, but neither the LTI nor the Army PEO have enough influence over concurrent acquisition efforts to effectively reduce power demand in the main electronics subsystems. An incentive structure would be one way to achieve innovation at the subsystem level. The shipping cost of batteries in the first Iraq conflict is estimated at more than $500 million, an indication of the savings that are possible in logistics alone. This includes only the cost of the logistical support, not of the batteries themselves. By reducing the average power demand by only 10 percent, a saving of $50 million could have been realized. This would be enough to develop five chips at $10 million apiece. The cost per soldier of providing batteries and related power source hardware, such as chargers, will be substantially higher as the OFW electronics suite is introduced. It is also possible to have reductions of much more than 10 percent. Finally, if the power requirements are reduced, the type of energy sources can be changed, which could also save costs. The committee therefore recommends that the Army should undertake a complete life-cycle cost analysis to determine the overall savings achievable by a substantial increase in development activity solely targeting power reductions (such as SoC design).

Soldier Communications

OFW is assuming the JTRS radio. Little information was available about the progress of JTRS design, but committee members familiar with software radio research suggested several areas for the Army to investigate for possible improvements in energy efficiency.

The JTRS has defined a standard based on the CORBA infrastructure, which was developed by the commercial sector for sharing software application modules over the Internet. While this has the advantage of defining a standardized interface, the power requirements associated with such an interface are likely to be very high, since it implicitly requires a software processor for the radio implementation. As shown in Figure 5-1, this approach will result in energy efficiency that is many orders of magnitude less than would otherwise be attainable. The commercial sector is also developing multiband and multistandard radios, but it uses software to reconfigure the hardware and is not considering approaches similar to the JTRS CORBA approach for its commercial applications.

The fully flexible JTRS radio will need an analog frontend that is tunable over a broad range of frequencies, from low megahertz to high gigahertz. This will require a considerable breakthrough in RF design and is potentially very inefficient with respect to power. The commercial approach is to use a number of duplicate RF chains, each of which is optimized for a single frequency band. It is suggested that careful consideration be given to how the analog portions of the radio are designed, since Moore's law technology scaling, which benefits the digital computation, will increase the power required for the analog circuitry.

To reduce the required transmit power, the soldier JTRS radio is being designed as part of a mesh network where each radio serves as a repeater to extend transmission coverage. However, care should be used in designing the ad hoc network protocols needed in such a network. Simulation studies have shown that the greedy approach—increasing transmission power only to the point of communication with at least one other node—leads to rapid depletion of the batteries of the nodes in the center of the formation as well as to added latency. Nodes in the center of the formation have to store and forward communications between nodes on the periphery, increasing local power demand and accumulating more latency for all messages. Recognizing and broadcasting over congested areas more evenly distributes power demand among all the nodes. Radio network simulations have been developed that accurately model radio power demand and these could be used to evaluate the energy efficiency of various protocols.

COMMERCIAL TRENDS

This section discusses new commercial trends and whether the trends highlighted in *Energy-Efficient Technologies* (NRC, 1997) are still valid and relevant to the Army. It also discusses the validity of projections made in that report and updates the original LW predictive model to reflect more recent goals and requirements.

Continuation of Moore's Law

The National Technology Roadmap for Semiconductors (NTRS) projections are still valid, as Moore's law is expected to hold true for at least another 8 years. The predictions of the 1994 NTRS tables regarding feature size and voltage reduction have been realized and even slightly exceeded (SIA, 1994). On the other hand, neither chip sizes nor the number of bits per chip on dynamic random access memory (DRAM) have grown at the projected rate, because lithography and manufacturing techniques at and below 0.1 micron are very expensive. Circuit development cost, particularly mask development, is rising with progressively smaller integration scales: from less than $1 million for micron-scale to $3 million for nanometer-scale.

The demand for ever denser circuitry has slowed, while demand for less complex application chips is growing exponentially. For example, a 90-nm complementary metal-oxide semiconductor (CMOS) can provide 8 to 64 MB/cm^2 (SRAM or DRAM) and 25 M logic gates. However one of the main sources of chip demand, the cell phone, typically requires only 2 M gates. Even most complex applications require on the order of 8-10 M gates. One of the largest commercial drivers is the personal computer (PC), but PC sales have declined since 1997. Industry is looking for new growth applications, but these new applications are likely to require fewer gates than cutting-edge technology. Even though application chips are less complex than state-of-the-art chips, costs and prices have still benefited from the feature size reduction. Decreasing demand for traditional masks in development may well combine with increasing demand for application chips to reduce costs.

Low-Power Electronics Technology

Since *Energy-Efficient Technologies* was written, there has been rapid progress in low-power technology, outstripping the Semiconductor Industry Association (SIA) roadmap, with increased performance and less power per circuit expected from scaling the transistor and wiring dimensions and to some degree the voltage. Since 1997, the energy efficiency of circuits has improved by a factor of at least 5. By one measure, the reduction in demand is greater than the improvement in rechargeable batteries, since time between recharges has only increased 20 percent. With increasing functionality, the net power demand for consumer electronics has remained essentially constant.

However, the exponential growth that has characterized improvements in commercial microprocessor performance may dramatically slow in the next few years. There are numerous barriers to progress after one or two more generations. The most relevant to low-power electronics are the various leakage currents that cause a passive power component, which is becoming increasingly significant compared to the active power.

It has been anticipated for some time that we cannot reduce the supply voltage much below 1 V for high-performance applications because the sub-threshold leakage current would go up greatly if the threshold voltage is scaled much below 0.3 V, as would be needed to maintain performance. For low-power applications at reduced performance, optimization studies for minimizing power at a given performance show that this threshold voltage needs to be *increased* as the voltage is reduced to maintain a balance between the static power and the reduced active power. Reducing the supply voltage down to about 0.5 V reduces the power more rapidly than the speed, with the result that energy/computation is improved, an important result for low-power electronics but made less palatable by the need to convert the voltage somewhat inefficiently from a higher voltage source.

Generally the energy-efficient design principles set forth in *Energy-Efficient Technologies* (NRC, 1997) are still good guidelines. The most important new trend is toward changes in technology and design to deal with the growth in static power relative to active switching power. This increased static power is from the increased transistor "off" current as the threshold voltage has been reduced, along with the operating power-supply voltage, to maintain performance growth. Static power also increases greatly with operating temperature. Less important, but rapidly becoming more so, tunneling currents in the gate insulator and at the drain-body junction are now limiting further transistor scaling, which requires thinner gate insulators and heavier body doping to maintain current trends. These limits are even more imposing for some applications such as static random access memory (SRAM) development, which is proving very difficult owing to threshold fluctuations in small-width devices.

As a result, several commercial chip design trends have developed:

- Technologies are being developed with multiple threshold voltages (V_t), allowing designers to sprinkle in low V_t transistors in performance-critical areas while using higher V_t transistors to save static power in less critical areas.
- The use of large switches in series with the power supply (header or footer devices) to turn off leakage current in inactive blocks of circuits. This is only effective after the stored energy in all the affected circuit capacitances leaks away.
- Reducing the power supply voltage to inactive circuits to reduce leakage. This is less effective than the use of large switches but allows the latched state to be retained.
- The use of adaptive body bias to raise V_t in inactive blocks or to compensate for V_t spreads due to process variations.

Except for multiple threshold voltages V_t (the first technique above), these techniques have been implemented mainly in low-power, battery-operated applications with significant standby time. Other trends aimed mainly at active power are (1) clock gating to turn off idle circuit blocks and (2) voltage-frequency scaling to reduce power (and energy/operation) when peak performance is not needed (Burd and Brodersen, 2002).

There are efforts in the industry to develop design approaches using nonvolatile memory, so as to have no leakage current in the memory elements themselves. EEPROM and FLASH, for example, are more storage devices than memory and can be used in special applications with relaxed requirements, such as write-cycle endurance. While nonvolatile memory is not likely to take the place of SRAM or DRAM embedded on processor chips, the nonvolatile memory used in the electronics and miniature hard drives for such products as digital cameras and personal digital assistants (PDAs) is very likely to find its way into some components of future soldier systems.

Changes in Commercial Development Trends

There is beginning to be a divergence between low-power personal computers and handheld devices (cell phones, PDAs, and the like). While computers emphasize higher processing speed with relative disregard for power leakages, handheld devices seek low-power implementations.

Radio hardware chips were universally proprietary in 1997, but many more chips are being made openly available today. This may represent an opportunity for the Army but would require the Army to evaluate the use of commercial waveforms for Army applications. Indeed, wireless LANs using 802.11 waveforms have already been used in the LW-Stryker generation for the soldier intercom.

Molecular electronics (e.g., nanotech applications such as individual transistors and carbon nanotube batteries) are viewed as follow-on technologies once the use of silicon plateaus. New materials, such as plastic semiconductors, are being developed for gates and displays and have the potential to one day be very inexpensive. The Army Institute for Soldier Nanotechnology is investigating the integration of circuitry into fabrics, so that uniforms for Future Warriors might well incorporate much of the electronic circuitry for LW functions.

Trends in Commercial Cell Phone Development

The development of dedicated applications such as cell phones emphasizes the system-on-a-chip approach, coupled with low-power architectures and aggressive powerdown strategies (see Chapter 6, "Power Management Approaches"). Power drain for traditional cell phones (voice only) is now considered to be under control. Digital logic originally required more power than analog, but both now require about the same amount. In the future, it is very likely the analog portion will become more dominant.

The evolution of multimedia on cell phones has renewed

the industry concern for power drain as a competitive factor. Soldiers will also require multimedia, but it is not clear how advances in consumer-electronics architectures will benefit the Army, since the applications are being optimized for consumer applications. Experience tells us that the Army will continue to use commercial developments more as a menu of prospective applications and functions, then engage a contractor to independently develop or integrate the most relevant functions.

Energy Efficiency of Integrated Circuits

In *Energy-Efficient Technologies* (NRC, 1997), projections were made of the evolving energy efficiency of processors and programmable digital devices. Those projections are still valid, but other factors that are critically important in determining energy efficiency were not taken into consideration.

An important aspect of evolving LW design has been the move toward an open architecture that can provide as much flexibility as possible. Among other things, such flexibility will make it possible to extend system capability into the future, to rapidly deploy prototypes, and—particularly in the communications area—to adapt to a wide variety of legacy systems. While it is clear that such a flexible, open architecture can be achieved, there is a wide range of alternatives that provide equivalent flexibility with greatly varying degrees of energy efficiency. Such alternatives range from the use of software on a conventional processor, through the use of interconnect reconfiguration, as in a field-programmable gate array (FPGA), to unique SoC circuitry.

The challenge is to come up with an SoC design that has sufficient generality to provide all the necessary modes but also specificity to the OFW problem domain. Solutions that are optimized for specific applications can have an energy efficiency that is several orders of magnitude greater than solutions providing for relatively unlimited flexibility. It is therefore critical that this cost of flexibility be understood, so that an informed decision can be made about how much flexibility is needed and to what degree specific tasks can be optimized for.

Energy Efficiency Metric

To quantify the cost of this flexibility, the committee considered the amount of energy, in millijoules,[1] required to execute an average operation. An operation is defined as a basic execution element, which for a software processor is an instruction and for other architectures would be an arithmetic or memory function, such as an add, subtract, or delay. The metric is millions of operations, MOP, per millijoule, or MOP/mJ. Energy efficiency can also be a more familiar ratio of rates: MOPS, which is million operations per second, divided by milliwatts, which is millijoules per second. Thus the energy efficiency metric is MOPS/mW, equivalent to MOP/mJ.

To see the relationship between this energy efficiency metric and the various levels of flexibility, a variety of designs will be compared that range from completely flexible, software-based processors (including both general-purpose processors and those optimized for digital signal processing) to inflexible designs using hardware dedicated to a single application, which it termed SoC solutions.

Metric Comparisons

In fixed-function designs, the operation count is straightforward, which is not the case for comparisons with processors that are flexible. In these cases different throughputs are possible, depending on the benchmark. When comparing architectures for different applications, the committee used the highest achievable throughput numbers.

To determine the state of the art for commercial circuits, the efficiency metrics for a number of chips chosen from the International Solid State Circuits Conferences from 1998 to 2002 are illustrated in Figure 5-1. (The chips selected had to be for a technology that ranged from 0.18 to 0.25 microns, and enough information had to be available to do a first-order technology scaling and to calculate the energy and area efficiencies) (IEEE, 1998; IEEE, 1999; IEEE, 2000; IEEE 2001; IEEE 2002). Though this is a relatively small sample of circuits, it is believed that the trends and relative relationships are accurate representations of the various architectures being compared because of the remarkable consistency of the results. Table 5-3 summarizes all the circuits that were used in the comparison. In the table, the designs are sorted according to their energy efficiency, and—very surprisingly—this sorting results in their being grouped into three basic architectural categories, which are differentiated by degree of flexibility. Chips 1-9 are general-purpose microprocessors and are fully flexible without any optimization for a given task. Chips 10-15 are software processors optimized for digital signal processing functions such as required by many of the OFW applications. Chips 16-20 are dedicated application SoCs, with very limited flexibility.

The medium-term estimates in *Energy-Efficient Technologies* (NRC, 1997, p. 137) predicted energy efficiencies of 2 MOPS/mW in 2001 for software programmable digital signal processors and 10 MOPS/mW for dedicated solutions. As seen now in Figure 5-1, these turn out to be conservative in comparison to the actual chips in 2001, which had efficiencies up to 10 MOPS/mW for the software digital signal processors and 100 MOPS/mW for dedicated designs. Figure 5-1 also illustrates that energy efficiency varies three to four orders of magnitude between the most flexible solutions and the most dedicated. It is not surprising that efficiency decreases as the flexibility is increased, but the flexibility is

[1]One millijoule (mJ) = 10^{-3} J.

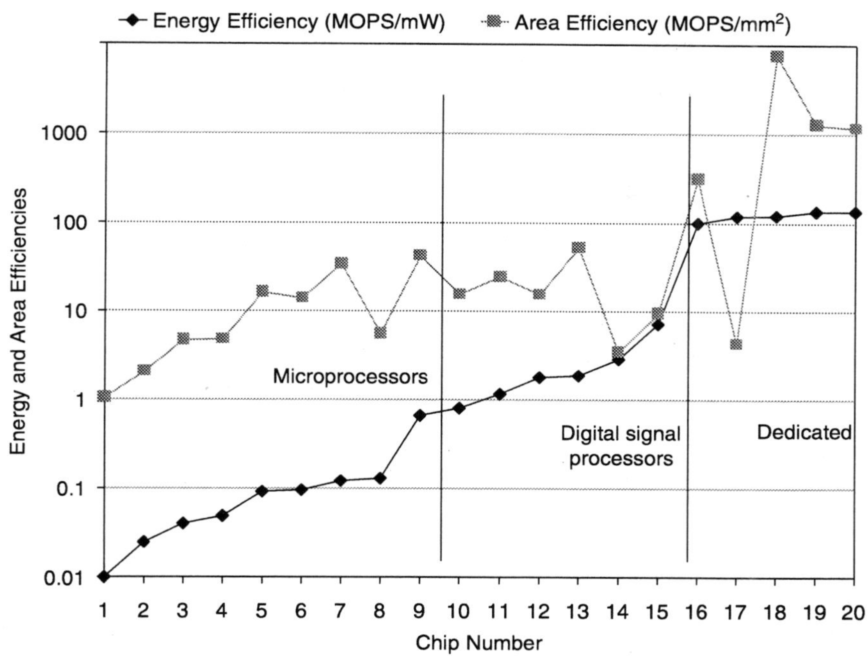

FIGURE 5-1 Energy and area efficiency of different chips from 1998 to 2002. Chips 1-9 are software programmable microprocessors, chips 10-15 are software-programmable digital signal processors, and 16-20 are dedicated designs. (See Table 5-3.)

TABLE 5-3 Description of Chips Used in the Analysis

Type	Chip No.	Year	Paper No.	Description
Software-programmable microprocessors	1	1997	10.3	µP—S/390
	2	2000	5.2	µP—PPC (SOI)
	3	1999	5.2	µP—G5
	4	2000	5.6	µP—G6
	5	2000	5.1	µP—Alpha
	6	1998	15.4	µP—P6
	7	1998	18.4	µP—Alpha
	8	1999	5.6	µP—PPC
	9	1998	18.6	µP—StrongARM
Software-programmable digital signal processors	10	2000	4.2	DSP—communications
	11	1998	18.1	DSP—graphics
	12	1998	18.2	DSP—multimedia
	13	2000	14.6	DSP—multimedia
	14	1998	18.3	DSP—multimedia
	15	2002	22.1	DSP—MPEG decoder
Dedicated designs	16	2001	21.2	SoC—encryption processor
	17	2000	14.5	SoC—hearing aid processor
	18	2000	4.7	SoC—FIR for disk read head
	19	1998	2.1	SoC—MPEG encoder
	20	2002	7.2	SoC—802.11a baseband

NOTE: µP, microprocessor; PPC, power personal computer; SOI, signal operating instructions; DSP, digital signal processing; MPEG, moving pictures expert group; SoC, system-on-a-chip; FIR, first-impressions report.

gained at enormous energy cost. With continued reductions in scale, from, say, 180 nm to 45 nm, an estimated 18-fold net advantage in efficiency can be expected.

For functions requiring a low number of operations to be executed (one such is the user interface), the energy cost of solutions providing high flexibility will not be a significant component of the overall system power demand. For functions with high processing rates, such as video processing and communications, solutions should be more dedicated to take advantage of the multiple orders-of-magnitude reductions in power that can be achieved. Combining applications that possess both low and high processing demands leads to the most generalized SoC approach, a chip design that provides software programmability where needed for functions that must have full flexibility and more dedicated, power-efficient solutions for high-performance signal processing. It is believed that more than an order of magnitude reduction in the power demand of the digital computation would be achievable in the OFW system if an SoC approach is taken.

On the other hand, it is clear that specification decisions that simply mandate a fully software-based system would result in crippling requirements from an energy-cost perspective, as appears to have happened in the JTRS program. For example, compatibility with multiple waveforms can be achieved by several strategies. Multiple dedicated radios could be placed onto a single chip, and since there is an area advantage similar to the power advantage for each dedicated radio using the SOC approach, a number of radios could easily be implemented in the same area that a software programmable solution would require. For the special case of radio designs, this is consistent with an approach that requires specialization in any case, because of the analog RF circuits that will require optimization if reasonable power levels are to be achieved. The energy efficiency of multiple dedicated radios on a single chip would easily be more than an order of magnitude better than that of a software-programmable solution.

Another development in the commercial arena that provides increased flexibility is use of reconfiguration as opposed to software programmability, such as used in field-programmable gate arrays (FPGAs). Computation is implemented on these chips using an architecture that is essentially the same approach as that used in dedicated SoCs, giving them an inherent advantage over a software processor-based solution. FPGAs are able to exploit the improvements in the underlying technology better than the software processors, so in the future it is likely that for high-performance computation requiring energy efficiency and full flexibility, the approach of choice will be based on reconfiguration.

At present, even though they are not optimized for energy efficiency, commercial FPGAs are still more efficient than software processors. The OFW soldier radio, for example, is being prototyped using FPGAs. An investment by the Army that would develop an energy-efficient, reconfigurable processor could achieve the dual goals of flexibility with reasonable energy efficiency. However, this solution will probably always be more than an order of magnitude less efficient than a more dedicated solution.

If all of the high-performance computation were integrated onto one OFW SoC chip the power demand of these functions could be reduced by more than an order of magnitude. Design of this chip using reconfigurable architectures could achieve these gains without compromising flexibility. In the OFW scenario, this flexibility could include radio and communication processors, Voice over Internet Protocol (VoIP) processing, as well as processing for video compression and decompression. For low-rate human interface processing, a software processor could be integrated onto the chip to provide additional flexibility to meet evolving future requirements. Chapter 6 discusses design concepts for such a Future Warrior system.

FINDINGS

The Army has come a very long way since *Energy-Efficient Technologies* (NRC, 1997) in understanding the soldier as a system and in taking appropriate actions that result from this understanding. Though in some cases there have been impressive reductions in the power demand of individual items, the reductions are being more than offset by the demands of new and more capable devices as well as the desire to have a highly flexible open architecture. Based on its observations of the overall evolution of the LW and OFW-ATD programs, the committee made six findings, which are discussed next.

Constraints on Reducing Power

For the LW systems the average power has been 20 W and the peaks have been 60 W over all three generations. The energy savings made possible by technology improvements have been traded for improvements in combat effectiveness as well as to allow the use of plug-and-play architecture to support future evolution. While the desire for such flexibility is understood, turning plug-and-play into a basic requirement comes at a high energy cost and will restrict the use of solutions that could reduce power demand by more than an order of magnitude.

Technology Time Horizon

The time horizon for OFW is too close. The LW program needs enough time to develop a SoC solution for the OFW and not be constrained to off-the-shelf component solutions. Increasing the development time horizon would allow the program to build on prior programs by evolving the SoC to meet new needs and requirements, similar to the successful approach taken for commercial cell phone evolution, in which each new generation is an enhancement of the last generation with new capabilities.

Life-Cycle Costs

The full life-cycle cost of providing power for soldier electronics is not being taken into account by the Army. The serious cost consequences of not using energy-efficient technology to design LW must be considered when determining the investments needed for reducing the power demand. The cost saving from requiring fewer batteries and other energy sources over the lifetime of the OFW system will more than pay for the development of highly optimized, low-energy solutions. For example, a 10 percent savings in power could be expected to reduce the number of batteries required by a comparable amount. This would have reduced the logistics cost of delivering batteries in Operation Iraqi Freedom, saving an estimated $50 million.

Soldier Communications

Power requirements for soldier communications are too great to ignore. As pointed out in *Energy-Efficient Technologies* (NRC, 1997), the requirements for soldier communications account for a considerable fraction of the overall energy consumption. The absence of reliable, more definitive estimates for the energy expected to be consumed by future OFW radios is thus of considerable concern. The solution being pursued by the OFW LTI is to use whatever radio is available in the time frame required for integration with OFW-ATD without any particular direct control of the radio design or its power requirements. In addition, the actual communication requirements and mission scenarios planned for JTRS are not necessarily synchronized with emerging requirements for network-centricity, and their lack of definition further obscures an already murky picture of what the OFW power demand will be.

In particular, the JTRS program, while certainly visionary in its goal of achieving compatibility with multiple communication waveforms, is, perhaps, overly ambitious. The software-defined radio on which the soldier radio is based will require advanced use of integrated circuit technology (highly integrated, mixed-signal SoC chip designs) as well as breakthroughs in protocols and architectures. Further, the design approach is unique to DOD, which effectively prevents the military from leveraging the gains in energy efficiency that are expected in commercial communications gear.

The OFW-ATD and LW program place considerable reliance on these JTRS developments with no guarantee that power reduction enjoys an equally high priority in the JTRS program. Without question, the power budgeted for communications is excessive considering the state of the art in communications electronics.

The power sources community understands well the inherent limitations to achieving large, short-term, step function improvements in the energy:weight ratio of power sources. However, on the demand side, the communications-electronics (circuits and systems) community continues to use traditional military approaches to circuit and systems design that are based on mechanical transport and modular interchangeability. These approaches lag well behind the capabilities of private industry and will prevent the Army from reducing energy consumption for soldiers who must communicate to survive on the battlefield.

Design Approaches

Use of application-specific integrated circuits (ASIC) and SoC design techniques are essential and could reduce power by more than an order of magnitude for digital computing and communications processing, making it negligible in comparison to analog sensor and display demand. There has been no effort in this direction in spite of the recommendations of the earlier study. Effort is also needed in reducing the power required by the analog portions of OFW electronics, particularly in communications devices.

Incentives for Reducing Power

The present focus on improving combat effectiveness may not result in net power reductions. While development, integration, and procurement contracts may contain goals for power, there are no financial (or other) incentives to make improvements beyond requirements. In fact, there is a disincentive to reduce power—namely, the potential for increasing risk and near-term cost.

The Army should turn the potential for logistics savings into an incentive. For example, a design saving 1 W could result in a savings of 864 Wh per LW system if the soldier participates in one 72-hr mission a month. Assuming 200 Wh/kg batteries, this would eliminate the need for almost 10 pounds of batteries per year. At $35,000 per ton to deliver supplies to a combat area, the transportation savings is over $100 per system per year. Assuming a 10-year system life and 1,000 soldiers, such a design that saves 1 W is worth $1 million in transportation savings alone.

It took an estimated $500 million to ship (not buy) the batteries used in the first Iraq conflict. By reducing average power demand 10 percent, a $50 million savings could be realized. This would easily cover the development cost of five chips at $10 million per chip or pay for several iterations of a custom ASIC design. If the cost of batteries and related power source hardware, such as chargers, envisioned for the OFW electronics suite is added to the calculation, the savings per soldier would be substantially greater than 10 percent. Perhaps most important, reductions in power demand may well reduce the complexity of energy sources needed and provide additional dollar savings.

Therefore, a reasonable recommendation is for the Army to perform its own life-cycle cost analysis before deciding what it can and cannot afford in the way of development costs. The committee believes the savings revealed by such an analysis would easily justify paying contractor incentives and increasing development activity on energy-efficient design approaches to future Land Warrior systems.

6

Future Warrior Design Concepts

Chapter 5 explained that the energy efficiency of circuits has improved at least fivefold since 1997. In the same period, system designers added new functionality to dismounted soldier applications with the expectation that existing power sources, or those being developed, would prove sufficient. The simple fact, however, is that the laws of physics and chemistry preclude the development of wearable power sources with the needed combination of energy, weight, and size that will not affect the soldier's agility. It is therefore imperative that the Army devote R&D effort to reducing power demand as it continues to develop and improve its power sources.

This chapter discusses design concepts for the Land Warrior (LW) systems to be borne by future warriors in the far term (2020 and beyond) and lays out a grand challenge for the Army to reduce overall power demand by an order of magnitude, from the 20-W regime now contemplated to 2 W or less. By adopting state-of-the-art (SOA) commercial design practices and incorporating power management technologies, peak power demand on energy sources can be reduced, increasing the combat effectiveness of individual soldiers and extending the duration of their missions. Aggressive power system designs tailored to the applications could take into account soldier modes of interaction and reduce power requirements for computation and communications, without affecting the effectiveness of the ensemble.

LOW-POWER SOLDIER SYSTEM

The numerous advantages of reducing power demand to 2 W from 20 W are obvious and undeniable. The committee believes that such a goal is attainable and should motivate the Army to expand its role in developing the soldier system. Perhaps most importantly, a 2-W system would not be dependent on developing Army-unique power systems. To illustrate, the committee reviewed and assessed power source technologies for a hypothetical regime that would include electronics applications demanding 2-W average and 5-W peak powers.

2-W Average with 5-W Peak

The total system mass of three batteries and two fuel cells that produce 2 W average power is shown for 24- and 72-hour missions in Figure 6-1. The battery technologies chosen for comparison are SOA primary Li/MnO_2 (280 Wh/kg, as extrapolated from Table D-1), secondary Li ion (170 Wh/kg, see Table D-2), and Li/$(CF)_x$ (820 Wh/kg, extrapolated from Table D-1). The specific energies used for Li/MnO_2, Li ion, and Li/$(CF)_x$ are reasonable projections for these chemistries but do not include a penalty for packaging multiple cells. For example, currently packaged LW batteries have the following characteristics: 195 Wh/kg for the 175-Wh LM11 Li/MnO_2 primary battery and 145 Wh/kg for the 135-Wh LI9 Li ion secondary battery (Brower, 2003). The points for Li$(CF)_x$ are based on a much larger cell, which might not yield this performance in the sizes and discharge rates required for these missions, even though improvements in the system are likely.

Data for other battery chemistries that are discussed in this chapter would fall within the range bounded by the Li ion and Li/$(CF)_x$ data. The only other technology discussed is the passive direct methanol fuel cell, which is at an advanced stage of development. The points are derived from data produced from fully packaged systems including fuel.

Several battery technologies meet the 2-W regime power requirement with a mass of less than 1 kg. For both 24- and 72-hr missions, it is clear that SOA primary and secondary batteries are the best choice and, with further improvement, might be the only choice. Given additional development time, passive direct methanol fuel cells (DMFCs) might yield performance comparable to that of some of the batteries. A possible basis for pursuing the DMFC rather than batteries would be logistics, not performance. Any

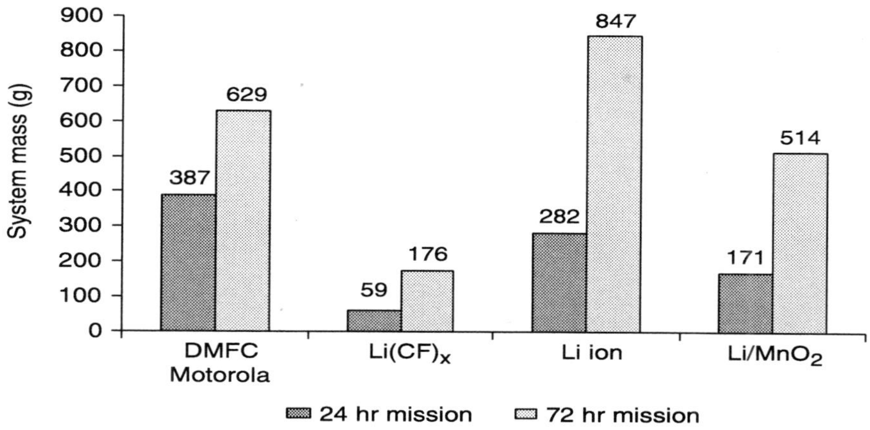

FIGURE 6-1 System mass of five energy sources producing 2 W average power for 24- and 72-hr missions.

battery with a specific energy exceeding 300 Wh/kg would exceed the performance of both DMFC and Li/MnO_2 and would be a good candidate.

As with the 20-W regime, if the same battery were used for average and peak power, the degree to which the peak power demand would degrade the specific energy would depend on the duty cycle. In the 2-W regime, a capacitor could be used for the 5-W peak if the pulse width is sufficiently short. An appropriately sized Li ion battery or a capacitor would most likely provide the 5-W peak power for the passive DMFC option, which would be optimized for the average power load.

At such low powers, advances in thermoelectric materials make them viable candidates for energy conversion systems. Thin film materials could enable very lightweight thermoelectric modules. Furthermore, they could be used in conjunction with catalytic combustors employing jet propellant 8 (JP-8) fuel, which has twice the energy content of methanol. However, even extremely optimistic energy efficiencies for a thermoelectric system are less than the 20 percent efficiency of a passive DMFC system. Other factors, such as system simplicity or size and weight, could make the thermoelectric option viable for some applications.

At the 2-W power level it is also possible that energy harvesting technologies could affect the overall weight and volume of the power source system. As with a DMFC system, the total energy harvesting system would have to perform significantly better than the battery it would replace. The total system weight for a given power produced and the often-intermittent nature of energy harvesting technology must be taken into account in any trade-off analysis. The significant benefit of using energy harvesting is the inexhaustible supply. This would free future warriors from concern about having no backup for dead batteries.

SYSTEM-LEVEL APPROACH

A 2-W soldier system would require a system-level approach to design that would consider both the energy consumers (sinks) and the power sources. Many techniques can be used to improve the energy-efficiency of a system, from the network level down to the physical level of the battery. At the network level, routing methods tailored to the power demanded by the network subsystem can improve power levels by 15 percent on average and reduce latency by 75 percent relative to methods that consider only the transmitted power.

At the boundary between the network and the processor levels, a computation can be performed locally or remotely depending on the relative performance of the local and remote system, the transmission bandwidth and power demand, and the network congestion. The largest power demand in a mobile computing system is for communication and computation. Techniques for reducing the energy usage in these areas include energy-aware network routing, balancing local and remote processing, and changing the central processing unit (CPU) speed dynamically.

Techniques for communication and computation cannot be studied in isolation: one technique for reducing communication energy usage is to perform more local processing, but this increases the amount of computation energy. Thus in the case of local vs. remote processing, the communication and computation subsystems must be considered together. For the system studied by Martin et al. (2003), energy can be conserved by remote processing any task that requires more than 1.4 milliseconds of processor time per kilobyte transferred. At the processor level, the main memory bandwidth has a significant effect on the relationship between performance and CPU frequency, which in turn determines the energy savings of dynamic CPU speed-setting.

The power sources must be considered as well. In particular, the energy delivered by a battery depends on the rate at which it is consumed. Consequently, reducing peak power can increase the battery life of the system by increasing the energy available. Large peak power has been shown to reduce the energy delivered by a battery by up to 40 percent. Electrochemical capacitors can be used to mitigate peak demands and improve battery life by up to 10 percent. The use of energy-aware operating system schedulers can reclaim even more of the battery capacity (Martin et al., 2003).

Table 6-1 summarizes mitigation techniques in key areas to improve energy efficiency. It lists improvements that could be realized by using a system approach toward mitigating energy issues associated with just the communications and computation functions of the Land Warrior.

In summary, greedy approaches to energy efficiency that consider subsystems in isolation will not be optimal. A system-level approach, one that considers energy consumers and power sources, is the proper method for examining energy efficiency in battery-powered computing systems.

POWER MANAGEMENT AND DISTRIBUTION

Land Warrior systems can be subdivided into four functional areas: displays, computing, sensors, and communications. Each functional area requires one or more power sources for electronics such as were listed in Table 5-1.

Distributed vs. Centralized

There are two basic approaches to power distribution: centralized and distributed. In centralized power distribution there is a central power source that is distributed by wires to the various power sinks. The centralized source may generate bulk power to distributed power regulators, which smooth out spikes and maintain voltage at the specified level for the various sinks (or for small, local, rechargeable batteries—this would allow various elements of the system to operate briefly if the centralized power source went down). If the system is small enough and the sinks have similar voltage requirements, the regulation may also be done at the centralized source. Since the energy density of batteries usually increases with battery size (owing to less packaging per unit volume), the centralized power option should result in higher energy density than the distributed power option and a lighter weight battery for the soldier.

The centralized power option should allow one to reduce the types of power sources needed. In theory, by coupling with DC-DC converters, it should be possible to use only one type of power source, with a secondary one of similar type as backup. Owing to the inefficiencies of DC-DC converters, there is a penalty of about 10 percent, but energy can be saved in the sink by keeping the voltage constant at the lowest possible level, particularly where the battery voltage changes greatly during discharge.

Additionally, central power will require that equipment be tethered to the central power source. The tethered equipment that would probably cause the most practical problems for the soldier is the weapon subsystem and the helmet-mounted heads-up display (HUD). The proposed OFW weapon subsystem has 14 components that draw power (Acharya, 2003). However, because a tethered weapon is prone to becoming entangled with protruding objects, its reliability should be studied in the field before implementation. Ultimately, the need to exchange an enormous amount of data via wires will dictate that the four functional areas of the OFW system be tethered together through wires. Consequently, one should be able to distribute the power from a central source through these tethered wires also. In other words, the unavailability of a wireless body LAN necessitates the use of tethered wires, which, in turn, dictates the centralized power approach. Advantages and disadvantages are summarized in Table 6-2.

In distributed power generation, the generation source, power regulator, and sinks are distributed around the body. There is no need for encumbering, tethering wires to connect

TABLE 6-1 Techniques for Mitigating Energy Issues in Key Land Warrior System Components and Improvements That Could Be Realized

Component	Mitigation Technique	Improvement
Power source		
Battery	Reduce peak draw	Up to 10% more available energy
Power sink		
Communications	Energy-aware network routing	Up to 50% fewer hops, 50% less energy
	Local processing	Delineation of local versus remote processing based on communications/processing cost
Computation	Remote processing	Delineation of local versus remote processing based on communications/processing cost
	Dynamic CPU speed setting	Prediction of idle time and active power within 5% of actual

NOTE: CPU, central processing unit.
SOURCE: Adapted from Martin et al., 2003.

TABLE 6-2 Advantages and Disadvantages of Centralized and Distributed Power Distribution for Use by the Dismounted Soldier

	Centralized Power	Distributed Power
Description	Centralized power generation, with either centralized power regulation or individualized power regulation associated with each sink.	Individualized power generation and power regulation associated with each sink.
Pros	Power generation more efficient at less total weight. Easier power generation maintenance. Logistics simplified owing to single power source type—no need for multiple different battery types.	Increased flexibility by having power associated directly to sinks. Fewer I^2R losses owing to shorter wire lengths. Survivability/reliability of system enhanced—equipment still functional if central power source lost.
Cons	Extra weight, due to wires for power distribution, and decrease in mobility, due to tethering sinks to the centralized power source (consideration moot if data must be transmitted from gun sight or helmet-mounted heads-up display with wires/cables anyway). Lacks power source redundancy/reliability gained with distributed systems.	Extra overhead weight associated with multiple power generation sites. Additional battery management overhead by soldier. May still require on-soldier battery recharging that could act as central power source.

the systems together (except, perhaps, to distribute data via wire for security). The obvious disadvantage of the distributed power option is that the dismounted soldier would have to carry varying quantities of different batteries, possibly increasing the weight carried for extended operations.

An alternative distributed battery scheme might utilize rechargeable distributed batteries that could be recharged on the soldier within a converter re-charger pouch. Hence the soldier would need (at a minimum) twice the number of secondary batteries as pieces of gear. This would require that each piece of gear be redesigned to accommodate a new battery geometry, which is harder than wiring an existing battery compartment to be powered via a cable, as is the case for a centralized power scheme.

Another consideration raised by the numerous types of batteries needed for the distributed power option is the likelihood that the soldier would need to replace batteries during operation, which could endanger his or her life in the midst of a firefight. There is also added weight for such things as battery containers and basic components for power regulation. A possible (but somewhat impractical) remedy would be to standardize the voltage range of batteries for all the energy-consuming devices so that only one type of battery is needed. However, this would increase the power budget of the OFW system.

The initial OFW concept combines the advantages and disadvantages of both centralized and decentralized power distribution. Many of the sensors have dedicated power sources, yet the sensors are tethered together using data cables. (Note that current USB hubs can routinely source 2.5 to 4 W per node, so that both limited power and data can be provided in a common connector/communications format. This protocol—as well as the 1394 FireWire protocol—is being studied for use by OFW). In addition, the different sensors may have different battery types, compounding logistics for the individual soldier and for supplying field units. For example, if there are 10 separate battery sources, each sized to the duration of a mission, a soldier might have to replace 10 batteries, possibly on 10 different occasions.

In summary, in the near term, the need to exchange an enormous amount of data dictates the use of tethered wires, thus favoring centralized power distribution and management. In the long term, when the wireless body LAN is available, it will enable more robust distributed power distribution. However, for the total OFW system power estimated, the power sources are still too heavy to allow for the distributed power option. More effort should also be devoted to simplifying the components of the weapon subsystem in order to reduce complexity and, consequently, the weight budget of the power sources needed.

Realistically, advanced power sources *and* aggressive power management will be needed to reduce the weight of the power sources, thus enabling the more robust distributed power option. As discussed in the following section, using power management in the design stage could reduce the total OFW system power to less than 2 W and enable deployment of commercial off-the-shelf power sources.

Power Management Design Approaches

This section discusses approaches toward reducing power demand by including power management concepts in the system design of LW systems. The most efficient and straightforward way to obtain energy savings is to power down any circuitry that is not being actively used. The power-down approach can be applied directly, or it can also form the basis for indirect techniques, such as in smart dust and asynchronous designs. Although soldier requirements

TABLE 6-3 Subsystems in Objective Force Warrior with Estimated Duty Cycle of 0.98 W

Function	Peak Power (W)
Communications processor and backplane	4.4
WLAN card	0.6
VoIP	0.6
Dead reckoning	0.25
Microphone/speaker	0.5
HIA	3.0
Weapon hub	1.0

NOTE: WLAN, wireless local area network; VoIP, Voice over Internet Protocol; HIA, high integration actuator.

SOURCE: Adapted from Erb, 2003.

are volatile, aggressive power-down—coupled with low standby power demand—can be directly applied to the OFW design.

One method of reducing the power for LW would be to utilize power-down technology for devices with heavy duty cycles. As depicted in Table 6-3, a number of OFW subsystems are almost continuously drawing peak power. The average power for these subsystems (7.986 W) is almost half (0.466) of the total system average power (17.149 W).

Similar heavily utilized subsystems also appear in Stryker. Table 6-4 shows that the average power for these subsystems (13.1 W) is more than two-thirds (0.672) of the total system average power (19.5 W). Two of the subsystems, the computer and the wireless local area network (WLAN) card, could use more aggressive power management protocols—for example, the WLAN cards can use a beacon technique wherein they awaken for 10 milliseconds every second. If any card has information to transmit, it would send a keep-awake message during the beacon period. The WLAN cards stay active until all the information has been exchanged. In times of little information exchange, the WLAN subsystem would have a duty cycle approaching 0.01. Likewise, the computer system and voice communications could be activated only when required. An onset-of-speech circuit, coupled with a 10-millisecond circular audio buffer, could wake up the other subsystems when there was audio activity. The buffer ensures that the initial part of the speech utterance is not clipped.

In the present OFW design concept, almost half of the average power demand comes from components with duty cycles of over 90 percent (see Tables 6-3 and 6-4). This is so in spite of the fact that functions such as voice communication could operate at duty cycles of less than 10 percent and network components that are commercial devices (such as the network hubs) could be designed to be actively powered off instead of simply left on all the time. An additional problem is that even when many of the components are put into standby, they continue to need significant power. Without design change, this problem will probably become worse, since the digital circuitry itself will have leakage currents that can only be controlled through intentional design strategies.

To accurately determine power source requirements, it is necessary to measure the actual duty cycles of the various components used in soldier operations. This measurement can be used to model the active, peak, and standby power of all the components, so that each power sink can be simulated. These models can be used in a full simulation of the dynamic operation that works with a source simulator (e.g., the White source simulator). This full simulation of OFW power sources and sinks should be used to make the investment decisions, which will have the most impact on developing hardware that substantially reduces power demand.

These and other simulations to resolve soldier power issues will require high-fidelity models and high-performance

TABLE 6-4 Subsystems in Stryker with Average/Peak Active Power Ration Greater Than 0.50 W

Function	Peak Active Power (W)	Fraction of Time Active	Average Power in All States (Standby, Active, Peak) (W)
Computer DRAM	2.7	0.833	2.25
Computer/master hub	0.7	0.29	0.37
Computer processing card	40	0.29	2.88
WLAN card	1.5	0.783	1.3
WLAN digital radio	0.17	0.813	0.142
WLAN antenna and power amplifier	1.63	0.813	1.39
Microphone/speaker	0.3	0.813	2.08
GPS	0.17	0.813	0.142
Dead reckoning	0.25	0.813	0.254
HIA	0.76	0.98	0.76
Laser energy detector	0.5	0.832	0.417

NOTE: DRAM, dynamic random access memory; WLAN, wireless local area network; GPS, Global Positioning System; HIA, high integration actuator.

computer capabilities. The Army should therefore consider taking advantage of the high-performance computing (HPC) assets at the two DOD major shared resource centers—the Army Research Laboratory (ARL) and the Engineer Research and Development Center (ERDC).

Smart Dust

Smart dust sensor networks show how savings can be realized using a power-down strategy and high levels of integration with system-on-a-chip (SoC) technology. Such networks are estimated to have lifetimes of more than 3 years from two AA batteries. This is accomplished using conventional hardware and a microprocessor that has an energy efficiency of 1.6 MIPS/mW, which is a typical value for software processors. The energy efficiency of the approach is gained almost entirely from the aggressive power-down strategy and protocols, which require the device to be active for only 1 or 2 percent of the time. This strategy is coupled with sleep modes, which need only microwatts of power to achieve the long unattended lifetime required for that application.

"Smart dust" is a term that was coined by Professor Kris Pister of Berkeley under a DARPA program to develop MEMS-based sensors and electronics, but this term is now also used broadly to describe sensing and transmission systems on a chip. The "dust" part of the program has not been realized, but the development of autonomous sensors and their role in networked systems may have a significant impact on the military. Early prototypes have shown potential for multiple years of operation, assuming that the data will be transmitted over relatively short ranges. Such short communication lengths assume that the devices are dispersed in aggregates and then use multihop transmission to transmit data to a base station with more power for longer-range transmission to a military installation or soldier. The power needed to transmit increases exponentially with the transmission distance, which is dramatically reduced when a multihop strategy is employed. This also is an important energy conserving concept.

The smart dust concept becomes impractical for truly dust-sized or micron-sized devices. Presently, these devices are realistic only if they are several cubic centimeters in size with relatively limited electronic sensing and transmitting functions. A concept even more challenging envisions mite-sized devices that would have computing capabilities for intelligent sensing plus the ability to move (crawl, fly, and/or climb)—such devices have been termed "cognitive arthropods." The "bugs" would require current on the order of 40 mA not only for transmission but also for mobility and would occupy a total space of 2-3 cubic centimeters, with 1 centimeter allocated for the energy source.

The proceedings of a workshop on cognitive arthropods (Main, 2003) suggested that approximately 30,000 Wh/L would be needed for these smart bugs to carry out the stated military goals. Batteries and fuel cells obviously cannot deliver this quantity of energy, leaving micronuclear or energy-harvesting devices as the only power source option. Sensors can be employed today for autonomous networks with limited sensing and transmission capabilities, but devices that are dust-sized are still well in the future and will require innovations in computing and cognitive capabilities as well as in energy sources. The direction of smart dust development is integration of the sensing, computing and communication functions into an SoC. This lends support to the committee recommendation to use SoC in developing soldier electronics and illustrates the importance of an aggressive power-down strategy to conserve energy.

Very Low Power Asynchronous Circuitry

Another way to optimize low-energy operation uses essentially the same strategy at the circuit level. Unlike clocked circuits, asynchronous circuits only turn on when notified that new data are available. This approach is still at the research stage, but it might be used for application-specific integrated circuit (ASIC) and system-on-a-chip (SoC) designs once it is fully developed.

Private companies have worked on asynchronous systems over the past three decades with varying results, but there is now a strong resurgence of interest because of the increasing power drain and power density in high-performance chips. Asynchronous system technology serves to reduce power drain by dealing with timing problems in circuits using sub-100-nanometer processes. The circuits operate only when a signal activates the circuit when new data are available.

The number of commercial applications requiring low-power design has increased dramatically, and there is growing interest in asynchronous systems for this reason as well. Large companies worldwide are now either using or exploring the use of asynchronous circuits for full or partially unclocked chips in order to significantly reduce the power wasted by having every switch receive clocking pulses whether it is involved in the calculation or not. The companies include Philips, Intel, Sun Microsystems, and many smaller companies. New designs have evolved that minimize the amount of handshaking required to synchronize the circuit elements in the absence of a common clocking impulse.

The use of asynchronous designs has recently been stimulated by Philips through its offer to share its tools, techniques, and design expertise and software with other companies. This is important, because the lack of design tools has been a major impediment to integrating advanced and sophisticated techniques into designs that could be used in soldier systems.

Asynchronous design especially makes sense for very large digital circuits with fast clocks, when the energy cost of maintaining global synchronization of the clocking network becomes prohibitive. However, this is a design space that currently has very little to do with soldier power problems. For digital signal processing, which is the dominant computation in the soldier system, asynchronous design is not so helpful, since there are real-time (synchronous) processing constraints that must be met. Also, the energy-efficient architectures for this kind of processing do not require high clock rates, so the energy cost of maintaining global synchronization is low. Finally, and most important, since the Army and its contractors do not even take advantage of conventional design strategies for integrated circuit digital design to save power, it would not make much sense to embark on a new strategy that is clearly at the early research level.

The best way to deal with the leakage issue is to work not with the fastest low-threshold (high-leakage) devices but with the low-power libraries, which means slower logic but lower leakage. Some designs (e.g., those for cell phones) use what are called foot (or sleep) switches, which put a switch in the supply lines to turn off the supplies during power-down periods. So the answer is not to go to the fastest clock rates but to use slower parallel architectures to do high-performance digital signal processing chores and avoid the power losses inherent to the clocking networks.

IMPACT OF SOLDIER INTERACTION ON ENERGY CONSUMPTION

Software designers have a rich variety of interface types to select from. Table 6-5 lists interface types in ascending order of required computing performance. As can be seen from the table, different interfaces require computing performance levels that differ by orders of magnitude.

TABLE 6-5 Computational Requirements to Support Different Forms of User Interfaces

Interface Type	Required Performance (MIPS)
Textual	1
Graphical user interface	10
Hand writing recognition	30
Speech recognition	150
Natural language understanding	1,000
Vision	10,000

NOTE: MIPS, million instructions per second.

The type of data to be exchanged must also be selected. Table 6-6 depicts four basic data types that can be used for transmitting a report. The simplest would be filling in a form by selecting values from a menu for each question in the form. To illustrate, assume that 100 different questions must be answered by selecting a single word from a menu for each question.

- *Text.* Assuming one word per question, an average of five characters per word, and eight bits per character, 4,000 bits of information would be generated.

TABLE 6-6 Sample Attributes of User Interfaces

Physical Interface		User Interface		Data		Compression		
Type	Relative Number of User Actions	Type	MIPS	Type	Bits	Type	Ratio	Number of Operations
Mechanical	3.50	Textual	1	100 words text	4,000	Textual	2	
Audio	1.25	Graphical user interface	10	60 s sound (at 2.4 kbs)	144,000	Video	30	8×10^6
		Speech recognition	150	Still picture (640 × 480) B&W, 16-level gray scale	1.23×10^6			
				Still picture (640 × 480) Color 24 bits	7.37×10^6			
				10 s video (640 × 480) B&W, 16-level gray scale	369×10^6			
				10 s video (640 × 480) Color 24 bits	$2{,}211 \times 10^6$			

NOTE: MIPS, million instructions per second; B&W, black and white.

- *Audio.* Assuming that the user files an audio report that requires 60 seconds to complete, sampling and encoding for audio require 2.4 kilobits per second.
- *Still picture.* Assuming a video graphics array (VGA) picture composed of 640 × 480 pixels with 16 levels of gray scale for black and white, the result is 1.23 million bits of data (e.g., 640 × 480 × 4). A color picture with 8 bits for each of the primary colors requires six times more data, or 7.38 million bits.
- *Video.* The report could also be filled with video clips. Assuming the same VGA quality as the still picture at 30 frames per second, a 10-second video clip requires 300 times more data than the corresponding black and white or color picture. The software designer can reduce the number of bits that need to be transmitted by applying compression algorithms. As shown in Table 6-6, a video frame can be compressed by a factor of 30 at the expense of eight million operations.

Once the user interface type and data type have been selected, the software designer can estimate the amount of energy required to support the interface and to transmit the data type. Once the total energy is known, the amount of battery weight needed for operation of the designed interface can be calculated.

Interface Design Example

Consider the design of the input interface using variations of the parameters shown in Table 6-6. Table 6-7 summarizes the battery-weight computation for a simple mechanical physical interface operating a textual software interface transmitting text, sound, still photographs, and color video clips. To provide contrast, an audio speech recognition interface for transmitting text and color video is also included.

For the textual input it assumed that one word can be selected from a menu every three seconds, yielding 300 seconds of operation for 100 words or 300 million operations per second (MOPS). It is assumed that the same report could be given with 60 seconds of audio clip, which requires 60 seconds of user interface usage. For both the still and the video it is assumed that the visual inputs can be captured with 30 seconds of interface usage.

The energy for computing can be estimated by dividing the total energy of the computing system by the million instructions per second (MIPS) performance rating from standard performance benchmarks. For the purposes of Table 6-7 we have assumed 0.1 MIPS/mW, which is typical of present day microprocessors (see Figure 5-1). In order to convert this energy into watt-hours we have to divide the energy per MIPS by 3,600, the number of seconds in an hour. Thus in Table 6-7 the energy for the textual interface with 100 words of text is

$$(300\text{ MOP})(10^{-4}\text{ W/MIPS})/(3{,}600\text{ sec/hr}) = 8.33 \times 10^{-4}\text{ Wh}$$

The energy to transmit data is the product of the number of bits to transmit times the energy per bit measured in watt-hours per bit. For this example we will assume a 1.5-GHz frequency, which would require 3.4×10^{-9} Wh to transmit a bit 1 kilometer in an outdoor environment with a moderate number of trees. Thus the amount of energy to send 4,000 bits is

$$(4 \times 10^3\text{ bits})(3.4 \times 10^{-9}\text{ Wh/bit}) = 1.36 \times 10^{-5}\text{ Wh}$$

Finally, the battery weight can be determined by dividing the total energy by the specific power of the battery measured in watt-hours per kilogram. Assuming 200 Wh/kg battery technology, the textual interface using 100 words of text as a data type would require

TABLE 6-7 Interactions Between User Interface and Data Types with Respect to Energy Required for Computing and Data Transmission

	User Interface			Computing		Transmitting			
Physical Interface	Type	Data Type	Interface (MOPS)	Compression (MOPS)	Energy (Wh)	Bits	Energy (Wh)	Total Energy (Wh)	Battery Weight (kg)
Mechanical	Textual	100 words text	300		8.3×10^{-4}	4×10^3	1.4×10^{-5}	8.5×10^{-4}	4.2×10^{-6}
	(1 MIPS)	60 s sound	60		1.7×10^{-4}	1.4×10^5	4.9×10^{-4}	6.6×10^{-4}	3.3×10^{-6}
		B&W still	30		8.3×10^{-5}	1.2×10^6	4.3×10^{-3}	4.4×10^{-3}	2.2×10^{-5}
		10 s color video	30		8.3×10^{-5}	2.2×10^9	7.53	7.53	3.8×10^{-2}
		10 s color video	30	80	6.8×10^{-3}	7.4×10^7	0.251	0.258	1.3×10^{-3}
Audio	Speech recognition (150 MIPS)	100 words text	45,000		0.125	4×10^3	1.36×10^{-5}	0.125	6.3×10^{-4}
		10 s color video	4,500	80	1.9×10^{-2}	7.4×10^7	0.251	0.27	1.4×10^{-3}

NOTE: MOPS, million operations per second; MIPS, million instructions per second; B&W, black and white.

$$(8.33 \times 10^{-4} + 1.36 \times 10^{-5})/200 = 4.24 \times 10^{-6} \text{ kg}$$

As can be seen from Table 6-7, the type of user interface and the type of data selected can have a dramatic impact on the energy consumption and consequent weight of the system. For example, a 10-second color video clip without compression would require 37.5 g of battery weight. Table 6-7 shows that the user interface design can affect energy consumption and battery weight by four orders of magnitude.

DESIGN GUIDELINES FOR WEARABILITY

Society has historically evolved its tools and products into more portable, mobile, and wearable form factors. Wearable implies the use of the human body as a support environment for the object. Clocks, radios, and telephones are examples of this trend. Computers are undergoing a similar evolution. Simply shrinking computing tools from the desktop paradigm to a more portable scale does not take advantage of a whole new context of use. While it is possible to miniaturize keyboards, human evolution has not kept pace by shrinking our fingers. There are minimal footprints beyond which objects become difficult to manipulate. The human anatomy introduces minimal and maximal dimensions that define the shape of wearable objects. The mobile context also defines dynamic interactions. Attempting to position a pointer on an icon while moving can be tedious and frustrating.

Wearability is defined as the interaction between the human body and the wearable object. Dynamic wearability includes the human body in motion. Design for wearability considers the physical shape of objects and their active relationship with the human form. Gemperle et al. (1998) explored history and cultures, including topics such as clothing, costumes, protective wearables, and carried devices. They studied physiology and biomechanics, movements of modern dancers and athletes. Also drawing upon their experience with over two dozen generations of wearable computers representing over a 100 person years of research, the results were codified into guidelines for designing wearable systems. These results are summarized in Table 6-8, which could be used to guide system designers of future warrior systems.

TABLE 6-8 Design-for-Wearability Attributes for Computers

Attribute Relating to:	Comment
Placement	Identify where the computer should be placed on the body. Issues include identifying areas of similar size across a population, areas of low movement/flexibility, and large surface areas.
Humanistic form language	The form of the object should work with the dynamic human form to ensure a comfortable fit. Principles include inside surface being concave to fit body, outside surface being convex to deflect objects, tapering sides to stabilize form on body, and radiusing edges and corners to provide soft form.
Human movement	Many elements make up a single human movement: mechanics of joints, shifting of flesh, and flexing and extending of muscles and tendons beneath the skin. Allowing for freedom of movement can be accomplished in one of two ways: by designing around the more active areas of the joints or by creating spaces on the wearable form into which the body can move.
Human perception of size	The brain perceives an aura around the body. Forms should stay within the wearer's intimate space, so that perceptually they become a part of the body. (The intimate space is between 0 and 5 inches off the body and varies with position on the body.)
Size variations	Wearables must be designed to fit many types of users. Allowing for size variations is achieved in two ways: (1) use of static anthropometric data, which detail point-to-point distances on different-sized bodies, and (2) consideration of human muscle and fat growth in three dimensions using solid rigid areas coupled with flexible areas.
Attachment	Comfortable attachment of a form can be created by wrapping the form around the body, rather than using single-point fastening systems such as clips or shoulder straps.
Contents	The system must have sufficient volume to house electronics, batteries, and so on, which in turn constrains the outer form.
Weight	The weight of a wearable should not hinder the body's movement or balance. The bulk of the wearable object weight should be close to the center of gravity of the human body, minimizing the weight that spreads to the extremities.
Accessibility	Before purchasing a wearable system, one should walk and move with the wearable object to test its comfort and accessibility.
Interaction	Passive and active sensory interaction with the wearable should be simple and intuitive.
Thermal	The body needs to breathe and is very sensitive to products that create, focus, or trap heat.
Aesthetics	Culture and context will dictate shapes, materials, textures, and colors that perceptually fit users and their environment.

SOURCES: Gemperle et al., 1998; and Siewiorek, 2002.

Long-term use of the wearable computers and Land Warrior ensembles may have unknown physiological effects on the human body. For soldiers, the combination of functions may also have unknown effects on combat effectiveness and performance. As such systems are used for longer periods of time, it will be important to test their effect on the wearer's body.

FINDINGS

Considering all levels in the system, power management of general-purpose computing functions can decrease power requirements by a factor of 2. Reducing peak power demand is especially important, because it increases the life of the energy source. Designing a system using aggressive techniques tailored to the application and to the user modes of interaction can reduce power requirements for computation and communication by several orders of magnitude. In turn, this will reduce weight required for the power sources and enable the system to utilize distributed versus centralized sources.

There are numerous ways to manage and reduce power using energy-efficient design techniques. R&D investment will be needed to enable the Army to evaluate the use of conventional architectures and logic and determine relative advantages of different architectures and circuits to employ for the analog and digital processing in soldier systems as well as to determine the appropriate levels of integration. Modern SoC technology must be demonstrated in soldier systems before its potential can be realized, and such technology will be essential to meet the grand challenge of a 2-W soldier system for future warriors.

7

Recommendations

This chapter summarizes key findings and provides the study recommendations. As requested by the statement of task, these include recommended priorities for investment in compact high-power and energy-dense source technologies for each of the regimes considered as well as specific recommendations in areas of hybrid systems, low-power electronics, power management and distribution, battlefield recharging, and predictive modeling.

The recommendations are presented in roughly the same order as requested by the task statement: beginning with findings in relevant technology areas (Recommendations 1-5), followed by findings for each of the power regimes (Recommendations 6-8), and ending with findings addressing specific topics (Recommendations 9-12). Recommendations 11 and 12 are considered overarching and are highlighted to show their overall importance.

POWER SOURCE TECHNOLOGIES

The committee formulated recommendations based on its evaluations of power source technologies in the regimes defined by the ARL/CECOM workshop. As detailed in Chapter 2, technology readiness levels (TRLs) were estimated for power solutions with the highest potential and then used as a basis for developing science and technology objectives and for making recommendations on technologies applicable to each target regime that are worthy of future Army investment.

Battery and Fuel-Cell Development

Batteries are the generic solution for soldier power. They will be an integral part of hybrid and stand-alone energy sources for the foreseeable future. The challenge is to make them smaller, lighter, cheaper, more energy-dense, more reliable, and with no sacrifice of safety. There is much commercial interest in achieving these ends, but these developments are designed for consumer electronics and are years away from being adapted as standards for the battlefield.

Recommendation 1: The Army should focus on batteries with a specific energy of 300 Wh/kg and higher for insertion into future versions of the Land Warrior (LW) ensemble. It should continue to promote and support innovative approaches to disposable and rechargeable batteries that can be adapted for military use. To select the best candidates for a given application, the Army should explore the trade-off space that exists between lifetime (measured in terms of charge-discharge cycles), specific power, specific energy, safety, and cost.

Fuel cells are the focus of intense interest by the military, primarily because of their potential as instantly "rechargeable" energy sources that can meet specific energy requirements for high electrical loads and long mission. Like metal-air batteries, fuel cells are air-breathing devices that cannot operate when submerged in water. Future acceptance of fuel cells on the battlefield will be determined to a great degree by logistics, because current prototypes are fueled by the nonstandard logistics fuels (methanol and hydrogen).

Recommendation 2: The Army should evaluate the applicability of small-scale, portable fuel processors capable of reforming the Army-standard fuels for use in proton exchange membrane (PEM) fuel cells or solid oxide fuel cells (SOFC). Scaling laws should be determined and cost/benefit analyses should be performed to determine whether there are power levels and/or mission durations that make such reformers an attractive alternative.

The Army must determine whether an alternative, nonstandard fuel source (such as methanol, hydrogen, or ammonia) is logistically acceptable. A proper analysis of trade-offs would permit decision makers to make an informed judgment on whether the operational advantages outweigh added logistics complexity and costs. Ideally, this would include testing in line units (even if only at the squad level) under representative field conditions. It would also save the Army money otherwise invested in research on fueled system alternatives that do not make logistical or operational sense.

Recommendation 3: The Army should immediately conduct a comprehensive and definitive analysis of the operational and logistical implications of fielding non-battery solutions as power sources for dismounted soldiers. This should include consideration of operational benefits, logistical limitations, and life-cycle costs, as well as considerations of safety and risk. It should develop models of competing energy sources, including fuel cell systems, and use them in simulations of battlefield operations. The data can then be combined with estimates of system costs to conduct cost/benefit analyses that would either support the consideration of non-standard-fueled fuel cell systems or eliminate them from consideration.

Small Engines

Several internal and external combustion engine prototypes have been demonstrated and show potential for military applications. Microturbine systems have not as yet demonstrated the capability to provide a net positive power output. Stirling engines use standard logistics fuel and could serve as a power source for battery rechargers or to meet anticipated requirements for high-demand microclimate cooling and exoskeletal applications. All small internal combustion engine systems now available have distinctive acoustic and heat signatures that would restrict their utility in combat. Stirling engines are inherently quiet but have significant thermal signatures.

Recommendation 4: The Army should adjust the focus of internal combustion engine development to demonstrate net power outputs and balance-of-plant systems appropriate to specific Army applications. Heavy emphasis should be placed on developing packaged systems with reduced heat and noise signatures. Once power output capabilities are demonstrated, the development should focus on improving system efficiencies.

Hybrid Power Systems

From a simple energetics point of view, hybrids offer enormous advantages for longer mission times. They also can provide a way to overcome the disadvantage of an air-breathing power source. Assuming that hybrids can be packaged to meet battlefield logistics and soldier operating requirements, they have the potential to replace batteries as the ultimate rechargeable energy source for soldier electronics. For fueled systems, efficient conversion at a modest 20 percent of the lower heating value (LHV) of the fuel leads to specific energy factors 2 to 5 times better than those of the best primary batteries.

Hybrids enable a system to be optimized for both high energy and high power demands. Some combinations, such as the battery-battery and battery-electrochemical capacitor hybrids, are air-independent and impervious to dust and moisture. Others that combine an air-breathing power source (e.g., metal-air battery, fuel cell, small engine) with a battery pose a problem for soldiers. To be acceptable, a fueled hybrid must be smart—that is, it must be capable of sensing and reacting to its environments so as to allow the unit to operate under water and to protect the unit from destruction.

Modeling is critical to the design of acceptable hybrid systems. The OFW-ATD program is collecting data to characterize Land Warrior power demands; it is possible that these data could also serve as a basis for modeling constructs to resolve soldier power issues.

Recommendation 5: The Army should refine duty-cycle estimates for the Land Warrior suite of electronics so as to enable the development of high-fidelity models incorporating soldier usage patterns and other details of interactions between power sources and soldier electronics. These estimates are essential for developing smart hybrid systems that can react to the environment for the future LW as well as for developing energy-efficient systems to meet unforeseen Army mission requirements.

Technologies for Target Regimes

While many commercial energy sources exist, they are motivated by a consumer market and are not developed in sizes commensurate with the broad spectrum of Army needs. The committee assessed technologies with high potential in each of the target regimes and determined science and technology (S&T) objectives for the near term (3-5 years), mid-term (5-10 years), and far term (beyond 10 years) based on a realistic appraisal of their current state of technology readiness. Table 7-1 lists the S&T objectives and indicates the relative risk (low, medium, or high) associated with each objective. Key research issues for each objective are enumerated in Chapter 2.

The committee was specifically requested in its task statement to select and prioritize power source alternatives in each of the target regimes.

20-W Average with 50-W Peak

Recommendation 6a: As its first priority in the 20-W target regime, the Army should support development of batteries with specific energies greater than 300 Wh/kg (e.g., Li/$(CF)_x$, Li/S, Li/air, C/air) in sizes commensurate with LW requirements.

TABLE 7-1 Science and Technology Objectives for the Near Term, Mid-Term, and Far Term, in Three Power Regimes

Power Regime	Near Term (3 to 5 years)	Mid-Term (5 to 10 years)	Far Term (beyond 10 years)
20 W average power	Develop batteries for the 24-hr mission with specific energies >300 Wh/kg.	Develop rapid start-up, compact solid oxide fuel cell (SOFC) systems operating on low-sulfur logistics fuel or surrogates.	Develop high-specific-energy, air-breathing battery system hybrids.
	Develop smart hybrid systems with high-energy and high-power batteries and/or electrochemical capacitors.	Develop complete small internal combustion and Stirling engine systems with low signatures operating on JP-8 or diesel fuels.	Develop microelectromechanical system components for power technologies.
	Develop generic modeling capabilities.		Develop SOFC systems that operate directly on high-sulfur and polyaromatic fuels.
	Develop efficient balance-of-plant components for small fuel cell systems.		
	Develop small fuel processors for logistics fuel, methanol, ammonia, and other viable fuels.		
	Develop and field-test direct methanol fuel cell (DMFC) hybrid systems.		
	Develop and field-test proton exchange membrane/hydrogen (PEM/H_2) systems.		
	Conduct battlefield-relevant safety testing of alternatives (H_2, MeOH, ammonia, JP-8, and Li batteries).		
100 W average power	Develop smart hybrid systems with small engines and fuel cells.	Develop small engines. Validate performance scaling laws. Assess reliability, failure modes.	Develop high-specific-energy, air-breathing batteries.
	Develop portable fuel processors for logistics fuel.	Develop SOFCs.	
	Evaluate DMFC and PEM systems for various specific missions.		
1 to 5 kW average power	Develop lightweight, efficient, 1- to 5-kW engines that operate on logistics fuel.	Integrate logistics fuel reformers with lightweight PEM fuel cells.	Develop high-capacity SOFCs and integrate them with logistics fuel reformers.
	Develop lightweight logistics fuel reformers.		

KEY: Relative risk: Low, ☐ ; Medium, ☐ ; High, ☐ .
NOTES: MeOH, methanol; JP-8, jet propellant 8; Li, lithium.

Recommendation 6b: The Army should develop smart hybrid systems capable of air-independent operation and the 50-W peak load. These hybrid systems must be developed with the aid of duty-cycle analysis and modeling. Key to this is an evaluation of the limits of battery-battery hybrid system performance as well as methods for packaging or sealing air-breathing hybrid systems.

Recommendation 6c: If the Army determines that a nonstandard fuel source is acceptable for battlefield use by dismounted soldiers (see Recommendation 2 above), it should develop PEM and SOFCs as complete systems with the hydrogen storage or generation subsystem yielding at least 6 percent by weight hydrogen, including all components. In this context, the Army should investigate

methods of reforming methanol, ammonia, butane, and liquid hydrocarbon fuels and should evaluate whether the development of direct methanol fuel cell (DMFC) systems would be less complex than fuel-processing approaches.

Recommendation 6d: As a final priority in the 20-W regime, and for the far term, the Army should develop and evaluate small engines that operate on standard logistics fuels.

100-W Average with 200-W Peak

Recommendation 7a: As its first priority in the 100-W target regime, the Army should develop smart hybrid systems capable of air-independent operation that can accommodate total energy requirements. The emphasis should be placed on fueled systems (small engines, fuel cells) capable of operating on standard logistics fuels.

Recommendation 7b: The Army should support development of high-specific-energy batteries for niche applications, such as laser designators.

1- to 5-kW Average

Recommendation 8a: As its top priority in the 1- to 5-kW regime, the Army should continue to develop lightweight engines with high specific power that operate on standard logistics fuels. It should investigate Stirling engines, as they are fuel versatile and offer significant acoustic signature reduction.

Recommendation 8b: For the 1- to 5- kW regime, the Army should develop the ability to process standard logistics fuels as needed for emerging high-specific-power PEM and solid oxide fuel cells.

SOLDIER SYSTEM ELECTRONICS

Considering OFW as the third generation, the average power has been estimated at 20 W and the peaks at 60 W for three successive generations of LW electronics. Power savings made possible by technology improvements in later electronics designs, primarily in the computer processors, have been traded for improved combat effectiveness as well as to allow the use of plug-and-play architecture to support future evolution. While the desire for such flexibility is understood, the approach comes at a high energy cost and restricts the use of more energy-efficient design solutions.

Energy-Efficient Technologies (NRC, 1997) determined that a Land Warrior averaging only 2 W would be possible if commercial design approaches, including system-on-a-chip (SoC) technology, could be applied to developing the soldier system. Use of SoC design techniques could reduce power by more than one order of magnitude for the digital computing and communications processing, making consumption negligible in comparison to that of analog sensors and displays. There has been no effort in this direction in spite of the recommendations of that earlier study. Special efforts are also needed to reduce the power demand of analog portions of the OFW electronics, particularly the communications devices.

The OFW-ATD Program has until the end of 2004 to integrate and demonstrate the third-generation LW. The length of the program, especially the technology time horizon, is too short to allow developing a SoC solution. The program is also constrained to using off-the-shelf components and cannot take advantage of true spiral development in evolving the soldier system. Consequently, it is unable to build upon the early LW program and evolve a SoC to meet new needs and requirements. In the evolution of commercial cell phones, the SoC approach has allowed each generation to enhance subsequent generations, bringing new capabilities at consistently lower cost in power. The committee determined that such a system would be easily powered by batteries already available. The cost savings from using fewer batteries would easily pay for any increases in program costs.

Both the LW acquisition program and the OFW-ATD program rely on separate Army programs to develop and acquire the component electronics. These other programs do not have the incentive to develop or procure electronics using commercially proven design approaches to reduce energy consumption, such as were described in Chapter 5.

Incentive is the key here. The Army buys things from companies oriented to the defense market but has provided these companies with few or no incentives to develop energy-efficient products. The committee believes that neither the LW acquisition program nor the OFW-ATD programs are large enough or have long enough development horizons to deal effectively with power issues. Power is a long-term concern that is drowned out by the Army's relatively near-term objectives.

As tempting as it may be for the Army to simply continue its use of outdated design techniques, a different strategy is required to design the equipment that the soldier must carry as compared with equipment that is carried on vehicles or other mobile or fixed platforms. Consider that there are important differences between what is required to design a smart cell phone and what was required to design an office telephone or a home computer. Just as cell phone users have special requirements, the soldier is a unique platform on which must be built a complex electronics system. For these reasons, it is important for the Army to increase its investment in Land Warrior electronics sufficient to begin a customized SoC approach to the development of these systems. In addition to reducing soldier energy needs, achieving energy efficiency for these electronics will resolve a myriad of problems now associated with the integration of disparate systems.

The Army acquisition system is impaired in its ability to focus on soldier power issues because it does not take into account the logistics costs of providing power on the battlefield when computing the true life-cycle costs of soldier electronics. The Army should take advantage of the new power-reduction designs and techniques that are well-known in commercial industry, especially in light of the high stakes involved for future soldiers on the battlefield.

Recommendation 9: The Army should make realistic estimates of the life-cycle cost, including reasonable logistics costs, of providing power on the battlefield and use such estimates in determining how much to invest in future Land Warrior design and development. Additional funding to extend the technology horizon of the program would enable a design solution that optimizes low-energy applications.

Power for Soldier Communications

Power and duty-cycle estimates for the LW soldier radio have not been refined for at least 5 years, even though communications technology has improved considerably. Wireless communications is the most power-hungry of soldier electronics applications and offers the potential for large reductions in energy requirements for the future warrior.

The importance of focusing on communications-electronics was emphasized in *Energy-Efficient Technologies* (NRC, 1997), but the Army has yet to pay attention. Five years later, the power performance planned by OFW-ATD for the Joint Tactical Radio System (JTRS) soldier radio is based on a rough equivalence with the MBITR radio, hardly the cutting edge of energy-efficient radios.

Power reduction has been given not nearly enough priority in the development of communications. Contracts that specify goals for power are not working. In fact, the added cost and risk of development serve as disincentives to reduce power demand. A thorough analysis of the communication solutions, mission scenarios, and resultant power demands is needed to determine if the power demand goals of the OFW program can be met.

Recommendation 10: The Army should make energy efficiency a first-order design parameter whenever specifying system performance parameters in its contracts. It should provide monetary incentives as needed to reduce power demand in all its procurements for soldier electronics, especially for communications.

OVERARCHING RECOMMENDATIONS

The OFW focus on increasing combat effectiveness rather than saving energy encourages trading off power savings achieved for new electronics. With no net reduction in power, this focus can undermine the benefits of system-of-systems design and contribute further to the chasm that exists between the state of the art in consumer electronics and in Army electronics.

Table 7-2 summarizes areas that are key to improving energy awareness and reducing power demand within the Land Warrior system. The first column lists major components of the system, the second column lists techniques for improving energy awareness, and the third column shows improvements that could be realized by using a system approach to mitigating energy issues associated with just the communications and computation functions of the Land Warrior.

To make progress toward providing adequate power for soldiers on the battlefield, the Army must shift its focus from providing energy to reducing energy demand, and it must do the hard job of developing a realistic mission profile. Recommendations based on these findings are considered of overarching importance in successfully confronting the issues of soldier power.

TABLE 7-2 Techniques for Mitigating Energy Issues in Key Land Warrior System Components and Improvements That Could Be Realized

Component	Mitigation Technique	Improvement
Power source		
Battery	Reduce peak draw	Up to 10% more available energy
Power sink		
Communications	Energy-aware network routing	Up to 50% fewer hops, 50% less energy
	Local processing	Delineation of local versus remote processing based on communications/processing cost
Computation	Remote processing	Delineation of local versus remote processing based on communications/processing cost
	Dynamic CPU speed setting	Prediction of idle time and active power within 5% of actual

NOTE: CPU, central processing unit.

SOURCE: Adapted from Martin et al., 2003.

Future Warrior Goal

The Army envisions a future uniform-and-electronics ensemble for the LW of the future. The committee believes that soldier electronics requiring a mere 2 W average power, 5 W peak power is attainable in the far term if the recommendations of this study are fully implemented. Using a 200 Wh/kg battery, available within the next few years, such a system could operate continuously for about 100 hr in a 1 kg package. However, concepts for powering the reduced needs of future soldiers should take advantage of likely reductions in the scale and distribution of power demand and consider options such as energy-harvesting technologies to provide reliable power at such low levels.

Recommendation 11: The Army should aim for a future soldier system capable of no more than 2-W average power, 5-W peak power. Achieving this will free the soldier from worries about power shortages on the battlefield and greatly enhance combat effectiveness.

Determining Energy Needs

Modeling has the potential to be a tool that saves time and money in developing efficient portable electronic systems if accurate system input can be supplied. Modeling can complement experimental data as it narrows down the parameters of optimization. The data ultimately need to be verified with experimental data, but the modeling can expedite the selection of a power source. Ideally, the military should develop and acquire new equipment based on recommendations and considerations from power sources modeling in order to maximize the lifetime of the equipment.

Substantial power reduction can be achieved through management techniques that power down unused components. Additionally, the power dissipation of components in standby mode should be reduced as much as possible; this will become increasingly important as silicon technology continues to lead to increased leakage currents. Actual measurements of the varying loads (rather than crude duty-cycle guesses) will allow simulating the dynamic operation of LW electronics in concert with a power source simulator.

At the highest level of simulation, given a range of mission scenarios, a suite of soldier equipment, and the size and makeup of combat teams, the Army should be able to determine optimum types, quantities, and distribution of power sources (and their fuel and recharging requirements). At the lowest level, the Army should be able to perform comprehensive analysis of every element and subelement in the entire system. Such an analysis must extend all the way from the leakage, clocking structure, and power-down capabilities on individual chips to the duty cycle on the laser designator, display, or radio, and everything in between. Engineers and scientists who are well versed in all of the modern technologies for very low power SoC design need to be sought out and used in this important effort to characterize the soldier requirement.

Full simulations of OFW power sources and sinks would also help to determine the directions that developments must take to have the most impact on power. While models based on experimental data can be used to expedite the proper selection and matching of power sources, higher order models could be used in simulations to tailor soldier applications to the most likely soldier modes of interaction, thus reducing power requirements for computation and communication. The Army has ready access to high-performance computing resources that are capable of supporting such important tasks, and such simulations can go a long way toward improving energy efficiency in military electronics.

Recommendation 12: The Army should develop a modeling capability for soldier equipment that includes power sources and also enables detailed simulation, verification, and analysis of power requirements for given operational parameters.

Ensuring adequate power for soldier systems is by no means a simple problem; if it were, the Army would not have asked the National Academies to do this study. It is a multidimensional challenge, and the solutions are found by considering not only energy sources but also energy sinks and energy management. The good news is that solutions exist in all regimes to satisfy known power requirements, and major breakthroughs in power/energy source technologies are not needed. To satisfy the needs of future warriors on the battlefield, the Army must move power to the forefront of considerations in developing and acquiring soldier electronics, especially communications. It also must invest in the means to analyze power requirements so as to take advantage of reductions that can only be achieved by efficient power management.

References

PUBLISHED

ASB (Army Science Board). 2001. The Objective Force Soldier/Soldier Team. Available at https://webportal.saalt.army.mil/sard-asb/ASBDownloads/OFS-ST.htm.

Atwater, T.B., P.J. Cygan, and F.C. Leung. 2000. Man portable power needs of the 21st century. I. Applications for the dismounted soldier. II. Enhanced capabilities through the use of hybrid power sources. Journal of Power Sources 91: 27-36

AUSRI (Auburn University Space Research Institute). 1994. Prospector VI Electric Actuation Workshop, March 21-23, 1994, Park City, Utah. Available from the Space Research Institute, Auburn University.

Batra, A., J. Balakrishnan, and A. Dabak. 2003. Multiband OFDM: Why It Wins for UWB. Texas Instruments. Available at http://www.commsdesign.com/story/ OEG20030624S0006. Last accessed on January 15, 2004.

Burd, T., and R.W. Brodersen. 2002. Energy Efficient Microprocessor Design. Kluwer Academic Press.

Dougal, R.A., S.Y. Liu, and R.E. White. 2002. Power and life extension of battery-ultracapacitor hybrids. IEEE Transactions on Components and Packaging Technologies 25(1): 120-131.

Fein, G.S. 2003. Manufacturers worked around-the-clock to replenish depleted stocks. National Defense Magazine, September. Available at http://www.nationaldefensemagazine.org/article.cfm?Id=1190. Last accessed on January 15, 2004.

Gemperle, F., C. Kasabach, J. Stivoric, M. Bauer, and R. Martin. 1998. Design for wearability. Proceedings of the Second IEEE International Symposium on Wearable Computers. Los Alamitos, California. Pp. 116-122. October.

Green, G., T. Bein, M. Pershbacher, D. Jordan, and J. Orens. 2002. Energy and Power Workshop for the Soldier. Final Report. Prepared for the Army Research Lab by Maritime Applied Physics and Berkley Research Associates. December 11.

IEEE. 1998. Digest of Technical Papers. International Solid-State Circuits Conference, 1998.

IEEE. 1999. Digest of Technical Papers. International Solid-State Circuits Conference, 1999.

IEEE. 2000. Digest of Technical Papers. International Solid-State Circuits Conference, 2000.

IEEE. 2001. Digest of Technical Papers. International Solid-State Circuits Conference, 2001.

IEEE. 2002. Digest of Technical Papers. International Solid-State Circuits Conference, 2002.

JASON. 2003. Portable Energy for the Dismounted Soldier. JSR-02-135. MITRE Corporation, McLean, Va. June 15.

Linden, H.D. 1995. Handbook of Batteries. 2nd ed. McGraw-Hill, New York.

Martin, T.L., D.P. Siewiorek, A. Smailagic, M. Bosworth, M. Ettus, and J. Warren. 2003. A case study of a system-level approach to power-aware computing. ACM Transactions on Embedded Systems 2(3): 255-276.

Martin, T., and D.P. Siewiorek. 2003. Non-ideal battery behavior and its impact on software design for wearable computers. IEEE Transactions on Computers, pp. 979-984. August.

Merryman and Hall. 1996. Chemical double-layer capacitor power source for electromechanical thrust vector control actuator. AIAA Journal of Propulsion and Power 12(1).

NRC (National Research Council). 1997. Energy-Efficient Technologies for the Dismounted Soldier. Washington, D.C.: National Academy Press.

Reilly, D., D.P. Siewiorek, and A. Smailagic. 2000. Power consumption and performance analysis of real-time speech translator smart module. Proceedings of the Fourth IEEE International Symposium on Wearable Computers, Atlanta, Georgia. Pp 25-32. October.

SIA (Semiconductor Industry Association). 1994. The National Technology Roadmap for Semiconductors. San Jose, Calif.: Semiconductor Industry Association.

Siewiorek, D. 2002. New frontiers of application design: Lessons learned from wearable and context aware computing. Communications of the ACM 45(12): 79-82.

UNPUBLISHED

Acharaya, V. 2003. Power Management for the Dismounted Soldier. Presentation by Vijay Acharya, U.S. Army Communications-Electronics Research, Development and Engineering Center, to a fact-finding group of the Committee on Soldier Power/Energy Systems. October 27.

Brower, W. 2003. Soldier Power Overview. Presentation by William Brower, Deputy Project Manager Soldier Power to the Committee on Soldier Power/Energy Systems, June and August.

Erb, M. 2003. OFW Power Management IPT. Presentation by Matthew Erb, Eagle Enterprises, to a fact-finding group of the Committee on Soldier Power/Energy Systems. October 8.

Graham, D., and S. Feldman. 2003. Power and Energy Systems IPT. Presentation by Dan Graham, General Dynamics Robotics Systems, and Scott Feldman, PM Soldier, during the General Dynamics Robotics Systems site visit on October 8.

Main, J. 2003. Exo-Skeleton Developments. Presentation by John Main, DARPA, to the Committee on Soldier Power/Energy Systems. Washington, D.C. June 19.

Masadi, J. 2003. Microclimate Cooling. Teleconference presentation by Roger Masadi, U.S. Army Soldier and Biological Chemical Command, Natick Soldier Center, to the Committee on Soldier Power/Energy Systems. October 30.

Pellegrino, J. 2003. Workshop on Energy and Power for the Soldier—Conclusions. Presentation by John Pellegrino, Army Research Laboratory, to the Committee on Soldier Power/Energy Systems. May 12.

Appendixes

Appendix A

Biographical Sketches of Committee Members

Patrick F. Flynn (NAE), *chair,* is retired vice president for research at Cummins Engine Company, Inc. He received his bachelor and master's degrees in agricultural engineering from the University of Minnesota, his MBA from Indiana University, and his PhD in mechanical engineering from the University of Wisconsin. Among other professional associations, Dr. Flynn was on the executive advisory board of the U.S. Army University Research Initiative and was on the advisory board for the Department of Energy's combustion research facility at Sandia National Laboratories. Dr. Flynn is a member of the Combustion Institute and a registered Professional Engineer in Indiana. He is a member of the NRC Board on Army Science and Technology (BAST) and served as a member of the Committee on Army after Next Logistics. He has expertise in mechanical engineering and integrated power systems.

Millard F. Rose, *vice chair,* is vice president for research at Radiance Technologies, Inc. He is former director of the Science Directorate at the National Aeronautics and Space Administration (NASA) Marshall Space Flight Center and also a former professor of electrical engineering and director of the Space Power Institute at Auburn University. He is author of well over 150 technical publications dealing with electromechanics, energy conversion, and environmental effects. Dr. Rose received his BS degree in physics from the University of Virginia and his MS and PhD in solid state science from Pennsylvania State University. He served as a member of the NRC Committee on Electric Power for the Dismounted Soldier and was chair of the NRC Committee on Unmanned Ground Vehicle Technology. Dr. Rose has expertise in electrical engineering and advanced power technology research and development. He is also an Institute of Electrical and Electronics Engineers (IEEE) fellow, an AIAA associate fellow, and a National Associate of the National Academies.

Robert W. Brodersen (NAE), a professor in the Department of Electrical Engineering and Computer Science at the University of California, Berkeley, has extensive experience with integrated circuit design and communications signal processing. He is a fellow of the IEEE and is a former member of the technical staff of the Texas Instruments Central Research Laboratory. Professor Brodersen is a co-director of the Berkeley Wireless Research Center, and he served on the NRC Committee on Electric Power for the Dismounted Soldier. His expertise is in the development of low-power electronics devices.

Elton J. Cairns is professor of chemical engineering at the University of California at Berkeley and was director of the Energy and Environment Division of the Lawrence Berkeley Laboratory from 1978 to 1996. He developed high-power-density secondary battery systems at the Argonne National Laboratory and was assistant head of the Electrochemistry Department at the General Motors Research Laboratories. He has served on several NRC committees, including the National Materials Advisory Board (NMAB) Committee on Electrochemical Aspects of Energy Conservation and its Committee on Production and Battery Materials Technology, and he was a member of the BAST Committee on Power for the Dismounted Soldier. Dr. Cairns has authored over 200 publications, and he holds 15 patents. His expertise is in battery and fuel cell research and development.

Huk Y. Cheh is vice president for technology at the Duracell Corporation. He received his BASc degree from the University of Ottawa and his PhD from the University of California at Berkeley, both in chemical engineering. He was on the technical staff at Bell Laboratories and headed the Department of Chemical Engineering and Applied Chemistry at Columbia University. Dr. Cheh has authored about 150 peer-reviewed papers and 7 patents. His expertise is in electrochemical systems and battery technology.

Walter L. Davis is vice president and director of the Advanced Consumer Systems Architecture Laboratory at the Motorola Corporation. As past director of Strategic Semiconductor Operations, Mr. Davis led a multiyear "war on power drain," which resulted in quantum improvements in the performance of Motorola paging and communications devices. He helped develop the first custom integrated circuits used in commercial radio equipment and holds over 75 issued patents in semiconductor and communications technology. Mr. Davis received his BSEE and MSEE degrees from the Illinois Institute of Technology. His expertise is in power distribution and management of low-power electronics.

Robert H. Dennard (NAE) is an IBM Fellow at the IBM Thomas J. Watson Research Center. He is an expert on solid state electronics and digital applications, particularly power issues and future technology trends. Dr. Dennard invented the one-transistor DRAM memory cell, coauthored the scaling rules that drive microelectronics, and has 30 issued U.S. patents. He received his BSEE and MSEE degrees from Southern Methodist University and his PhD from the Carnegie Institute of Technology. He served as a reviewer for the NRC report *Energy-Efficient Technologies for the Dismounted Soldier*. Dr. Dennard has expertise in power distribution and management for low-power electronics devices.

Paul E. Funk is director of the Institute for Advanced Technology at the University of Texas in Austin and former vice president of General Dynamics Land Systems. He is a retired Army lieutenant general who served as a division commander in the Gulf War, commander of the Third Armored Corps, and commander of the National Training Center. Dr. Funk received his EdD in education and training from Montana State University and has served on both the Defense and Army Science Boards. He has expertise in land combat, command and control systems, and military operations and requirements.

Robert J. Nowak is a consultant and former program manager at the Defense Advanced Research Projects Agency (DARPA) and the Office of Naval Research. He has directed and supported research in fuel cells, batteries, capacitors, energy harvesting, fuel processing, thermal energy conversion, microengines, hydrogen storage, biofuel cells, sonoluminescence, and biomolecular motors. Dr. Nowak initiated the DARPA Palm Power program, which focuses on portable power research and development for a variety of DOD missions. He received his BA and MS degrees in chemistry from Oakland University and his PhD degree in chemistry from the University of Cincinnati, and he was selected as NRC postdoctoral fellow at the Naval Research Laboratory in 1979. Dr. Nowak's expertise is in power and energy system technologies.

Jeffrey A. Schmidt is staff consultant and lead fuel cell technologist at Ball Aerospace Company. He was lead systems engineer on the SNORKLER program that developed the first successful man-portable, continuous-power PEM fuel cell for the DOD, and he has been lead integrator of fuel cell technologies for the DARPA Palm Power program. Dr. Schmidt received his BS degree in chemistry from North Dakota State University and his PhD in physical chemistry from Florida State University. He has expertise in the development of fuel-cell technologies and the integration of hybrid fuel cell systems.

Daniel P. Siewiorek (NAE) is Buhl University Professor in the School of Computer Science and the CIT Department of Electrical and Computer Engineering and director of the Human Computer Interaction Institute at Carnegie Mellon University. He has designed multiprocessor computer systems and authored textbooks on parallel processing, computer architectures, and reliable computing. He is a former consulting engineer for the Digital Equipment Corporation and the Naval Research Laboratory, was elected a fellow of the IEEE for contributions to the design of modular computing systems, and served on the BAST Committee on Electric Power for the Dismounted Soldier. Dr. Siewiorek's expertise is in assessing power distribution requirements for soldier-portable computer systems.

Karen Swider Lyons is materials engineer at the Naval Research Laboratory and former program assistant for the Office of Naval Research. She is principal investigator on distributed microbatteries for next-generation electronics devices and the development of catalysts for low-cost proton exchange membrane fuel cells. She is also technical advisor to the DARPA Palm Power program on portable batteries and fuel cells. Dr. Lyons received her BS in chemistry from Haverford College and her PhD in materials science and engineering from the University of Pennsylvania. She has expertise in portable power system technologies.

Enoch Wang is a program management engineer at the Central Intelligence Agency. He was formerly program scientist at the Duracell Research Center, where he conducted research for over 10 years, including 6 years of extensive experience with lithium ion materials. Dr. Wang received his BA in chemistry, his BS in chemical engineering, and his PhD in inorganic chemistry, all from Rutgers University. He is an active reviewer for the *Journal of the Electrochemical Society* and an officer in the U.S. Army Reserve. Dr. Wang has expertise in battery chemistries and technologies.

Donald P. Whalen is a consultant for Cypress International Corporation and retired U.S. Army brigadier general. While on active duty he served in the Office of the Deputy Chief of Staff for Research, Development and Acquisition, as program manager for acquisition of the Bradley Fighting

Vehicle, and as professor of electrical engineering at the U.S. Military Academy. General Whalen has served on the Army Science Board. He received his BS from the U.S. Military Academy and his MSEE from Purdue University. He has expertise in military operations and system research, development, and acquisition.

Appendix B

Committee Meetings and Other Activities

MEETINGS

First Committee Meeting, May 12-13, 2003, Washington, D.C.

Army S&T in Soldier Power and Sponsor Expectations
A. Michael Andrews, Deputy Assistant Secretary of the Army (Research and Technology)

Army Workshop on Power/Energy Technologies
John Pellegrino, Army Research Laboratory

DARPA Palm Power Program
Robert Nowak, Committee on Soldier Power/Energy Systems

Second Committee Meeting, June 19-20, 2003, Washington, D.C.

Exo-Skeleton Developments
John Main, Defense Advanced Research Projects Agency

Army Collaborative Technology Alliance on Power and Energy
John Hopkins, Army Research Laboratory

Land Warrior Power
Steve Slane, Army Communication and Electronics Command

Power Management Developments
Steve Slane, Army Communication and Electronics Command

Land Warrior Demonstration
Ross Guckert, Director, System Integration for PEO-Soldier
William Brower, Deputy Project Manager–Soldier Warrior
SFC Tony Husen, U.S. Army PEO-Soldier

Soldier Power Overview
William Brower, Deputy Project Manager–Soldier Warrior

U.S. Army Research, Development, and Engineering Command, Integrated Product Team (RDECOM IPT), Power and Energy Roadmap
Edward Shaffer, Army Research Laboratory

Third Committee Meeting, August 19-20, 2003, Washington, D.C.

AFRL Power Initiatives for Special Operations
Capt. David Pfaler, Air Force Research Laboratory

Power Aware Computers and Communications
Robert Graybill, Defense Advanced Research Projects Agency

Sensor Electronics
David Randall, CECOM Night Vision and Electronic Sensors Directorate

Army Research Initiatives
Richard Paur, Army Research Office

Fourth Committee Meeting, December 4-5, 2003, Irvine, California

SITE VISITS

Fuel Cells and MURI, University of California, Santa Barbara, August 4-5, 2003
Small Engines, University of Michigan, Ann Arbor, September 10, 2003
General Dynamics Robotics Systems, Owings Mill, Maryland, October 8, 2003

Microturbine Technologies, Massachusetts Institute of Technology, Cambridge, October 17, 2003

CONFERENCE CALLS

Soldier Power Overview Update, September 3, 2003
Predictive Modeling, September 5, 2003
Power Management, October 27, 2003
Microclimate Cooling, October 30, 2003

Appendix C

Measures of Performance

This appendix defines the standard measures of performance that were used by the committee to assess and compare power and energy source technologies. It describes thermodynamic limits of performance and derives weight limit criteria for fuel cell and small engine systems based on the system efficiencies of fuel and conversion systems.

DEFINITIONS OF TERMS

Power is the rate of doing work—or, alternatively, the energy or work produced or consumed per unit time—and is quantified in this report in watts (W). In the meter-kilogram-seconds (MKS) system of units, 1 W is equal to 1 joule (J) of energy per second (1 J/s or 1 kg–m^2/s^2), and in the British engineering system, 1 W is equal to 0.0013 horsepower (hp) (or, 746 W = 1 hp). Conveniently, 1 W is also equal to the electrical power dissipated when 1 ampere (A) of electrical current is carried through a 1-ohm resistive load, producing a voltage drop of 1 volt (V).

Power can take many forms, including electrical power, mechanical power, photonic power (electromagnetic radiation, such as radio waves or light), acoustic power, and thermal power. Electrical power, P_{elec}, in watts is simply the product of current, I, in amperes multiplied by the potential drop across the load, V, in volts:

$$P_{elec} = I \times V$$

The heat generation rate, P_{heat}, is also expressed in watts as the time rate of change (derivative) of heat, Q, in joules, over time, t, in seconds:

$$P_{heat} = dQ/dt$$

Energy is the amount of work done and can be calculated from the time integral of the power:

$$E = \int P(t)dt$$

The amount of energy can be quantified with the watt-hour (Wh), with 1 Wh of energy being equal to 1 W of average power integrated over a 1-hr period; it is equal to 3,600 J in the MKS system of units.

Capacity defines quantity of charge and is measured in units of ampere-hours or amp-hours (Ah). By definition, an ampere has units of 1 coulomb (C) of charge per second, so that 1 Ah is equal to 3,600 C. Capacity can be used to determine the coulombic efficiency of charging and discharging in batteries and the amount of fuel utilized by a fuel cell or engine.

Fuel consumption is quantified by the amount of fuel, in grams, used to produce a kilowatt-hour (kWh) of energy by a fuel conversion device. Grams per kilowatt-hour (g/kWh) is a standard and convenient metric for the fuel consumption of engines.

Wh is a measure of energy, but because the discharge voltage of the battery or operating voltage of the fuel cell is usually not constant, Wh, or energy from the power source, is not an accurate measure of the charge remaining in a battery or the fuel consumed by a fuel cell. The efficiency of batteries and fuel cells varies with current, so higher current (higher power) from a portable power source results in lower voltages. This point is illustrated by the curves in Figure C-1. Three discharge rates are shown, with each curve displaying voltages corresponding to the nominal capacity of the cell (in Ah) multiplied by the discharge factor (e.g., 0.2), the result being in units of amperes. Notice that the higher rates of discharge result in lower cell voltages and slightly lower capacities. These same features are present for the other charging curves, with higher recharge voltages being required at higher charging rates.

Specific power and power density are power per unit weight or volume of the system and are usually expressed with units of W/kg and W/L, respectively. Military systems

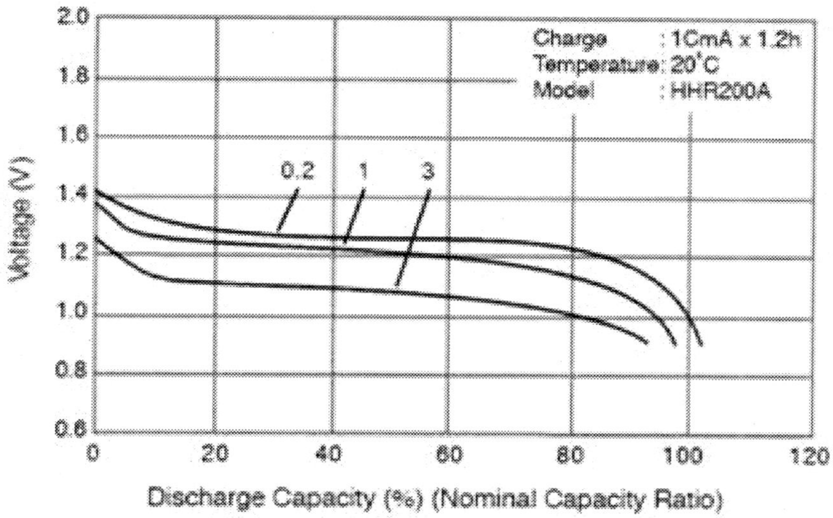

FIGURE C-1 The capacity of a battery changes with the rate of discharge. SOURCE: Panasonic, 2003.

typically focus on specific power, but the power density can be as important or more important for portable electronics, especially when a certain form factor is required. Because power is a function of the current drawn from the system, the power is often defined as the maximum power that can be delivered by the system. Alternatively, power can be specified at the practical operating current of the source. The specific power and power density of batteries are typically measured with respect to their full, packaged system. Fuel cells and engines can be defined by the power and weight of their conversion system only or of the conversion system and fuel combination.

Specific energy and energy density are the energy per unit weight or volume of the system. The most common units for portable power devices are Wh/kg and Wh/L. The amount of energy that is generated by a system is affected by operating conditions, as discussed above, so there is often a practical range in the reported values of the specific energy and energy density of battery chemistries. For clarity, the specific energy of systems is often reported as a single value, but it should be understood that, in practice, a variation should be expected.

The weight and volume term for batteries is measured for the full packaged system, and that for fuel cells and batteries includes the conversion system, plus the fuel and fuel tank, and relevant auxiliary components. Note that a fuel cell or engine with no fuel has no energy, so to accurately estimate these terms the amount of fuel consumed over a period of time must be known. Furthermore, when determining the specific energy and energy density of a developmental power source, care should be taken to qualify exactly what is included in the weight and volume terms, as the developer will not want to be saddled in the future with unforeseen components that add to the weight and volume of the system and make it less attractive. Additional weight might also be necessary for ruggedizing systems for military use.

Classes of energy storage/conversion systems are frequently compared by plotting the log of specific power as a function of the log of specific energy—such a plot is referred to as a Ragone plot (pronounced rah-GO-knee). The technique is used to compare classes of technologies and chemistries and can also be used to show how specific power and energy interrelate for a given battery or energy converter/fuel solution as it is discharged at different rates. Ragone plots are useful when trying to select a power source for high energy or high power. Figure C-2 shows an example of such a plot for several rechargeable batteries and an internal combustion engine. Note that the energy of a battery is inversely proportional to the power, so draining a battery quickly (at high current and, thus, high power) tends to lower the specific energy and draining it slowly (at low power) leads to higher energy densities. Care should be taken when evaluating batteries to discern whether the specific power and energy were determined at the same current, or whether they are the best-case scenario on the Ragone plot.

Efficiency

The efficiency, η, of an energy conversion or storage device is a function of both thermodynamics and engineering and determines the system energy and thermal signature. Systems can be described by either their thermal or electrical efficiency. The thermal efficiency, η_{th}, of an energy conversion device is defined as the amount of useful energy

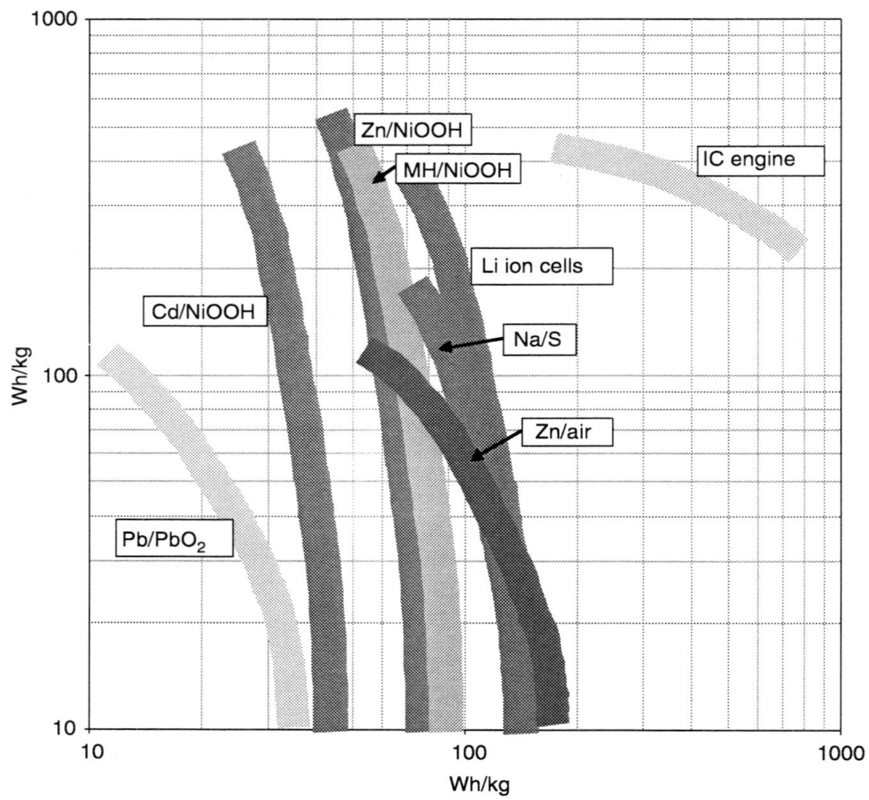

FIGURE C-2 Ragone plot comparing the specific energy vs. specific power of various batteries and of an internal combustion engine. SOURCE: Cairns, 2004.

produced relative to the change in the amount of stored chemical energy.

Fuel Cells

For a fuel cell, these parameters are described by the Gibbs free energy, G, and the enthalpy, H, of the system and fuel. The maximum thermal efficiency for such an electrochemical system is the ratio of the Gibbs free energy for the reaction to the enthalpy change for the reaction:

$$\eta_{th(max)} = \Delta G/\Delta H$$

The standard free energy change of a hydrogen-fueled system is typically calculated from the reaction of gaseous hydrogen to form liquid water:

$$H_2\ (g) + \tfrac{1}{2}O_2 \rightarrow H_2O\ (l)$$

At room temperature, this chemical energy of the system, ΔH, is 285.8 kJ/mole and the free energy for useful work, ΔG, is 237.1 kJ/mol, so the thermal efficiency of an ideal fuel cell operating reversibly on pure hydrogen and oxygen at standard conditions would be

$$\eta_{th} = \Delta G/\Delta H = 237.1/285.8 = 0.83$$

The values needed to calculate enthalpy and free energy of fuel cells reactions can be easily obtained from sources such as the JANAF Thermochemical Tables (Chase, 1986).

The efficiency of a real operating fuel cell is calculated from the actual vs. ideal voltage of the cell. The ideal (reversible) voltage of an H_2/O_2 fuel cell under no load at room temperature and pressure is 1.229 V when the product is liquid water (with a higher heating value [HHV]) and 1.18 V when the product is gaseous water (with a lower heating value [LHV]). Thus, the thermal efficiency of a fuel cell operating at voltage V_a at room temperature and utilizing all of the fuel, according to the last reaction above, is calculated from the equation

$$\eta_{th} = 0.83 \times V_a/1.229$$

Therefore, a fuel cell that produces liquid water operating at 0.8 V has an ideal thermal efficiency of 54 percent, while a fuel cell operating at 0.6 V has one of 40 percent. Of course, practical fuel cells do not usually consume all of the fuel supplied, and some leaves the system unreacted. This unreacted fuel needs to be taken into account in the efficiency calculation. The overall thermal efficiency would then be

$$\eta_{th} = V_a/1.229 \times U \times \Delta G/\Delta H$$

where U is the fraction of the fuel utilized electrochemically.

For solid oxide fuel cells (SOFCs) running on a hydrocarbon fuel, the parameters must be adjusted for temperature and the reactions for hydrocarbon oxidation. The ideal cell voltage of SOFCs is lower than that of proton exchange membrane (PEM) systems because voltage decreases with increasing temperature, but the LHV of gaseous water (1.18 V) improves the theoretical cell efficiency. Fuel cells do not achieve their theoretical efficiency because of ohmic losses within the cell due to materials resistance and polarization losses of the electrochemical reactions.

Small Engines

For engines, thermal efficiency is the ratio of the net work done, W_{net}, to the heat absorbed during a cycle. The amount of work is given by the net amount of heat converted into work by the engine, or the difference between the heat absorbed from the hot zone, Q_h, and rejected to the cool zone, Q_c, or sink:

$$\eta_{th} = W/Q_h = (Q_h - Q_c)/Q_h = 1 - Q_c/Q_h$$

For a Carnot cycle, the ratio of heat absorbed/rejected is equal to the ratio of temperature absorbed/rejected, resulting in the following equation:

$$\eta_{th(Carnot)} = 1 - T_c/T_h$$

Batteries

The same concepts apply to batteries, but here, thermal efficiency is not normally used. The efficiency of a rechargeable battery is simply the ratio of the energy obtained during discharge to the energy used for recharge. The ratio of all reactants that participate in electrochemical (faradaic) reaction to all the reactants present within a storage battery is referred to as the "utilization."

A rechargeable lithium battery operating at a nominal 3.7 V at rated power is operating with an efficiency of 88 percent, assuming an average charging voltage of 4.2 V. The high energy conversion efficiency of batteries adds to their value because it significantly reduces their thermal signature. The thermal energy released during operation of a battery is the difference between the enthalpy change for the cell reaction and the electrical energy produced by the battery. The thermal energy effect can be either a release of heat or the absorption of heat. The latter is rather unusual.

System Efficiency

The overall system efficiency, η_S, is a function of the thermal or electric efficiency of the power source and the balance-of-plant (BOP) efficiency, η_{BOP}, and fuel utilization, μ_f.

- For a fuel cell, overall system efficiency is given by $\eta_S = \mu_f \times \eta_{BOP} \times \eta_{ec}$
- And for a heat engine, system efficiency is given by $\eta_S = \mu_f \times \eta_{BOP} \times \eta_{he}$
- For a rechargeable battery, the overall efficiency is described as

$$\eta_S = \text{Energy delivered during discharge/Energy used during recharge}$$

The BOP efficiency includes the electrical and thermal penalties for the system, among them the electrical power needed to run air and fuel pumps and the thermal losses or cooling requirements. For low-temperature hydrogen PEM fuel cell systems, BOP efficiency is typically assumed to be better than 95 percent.

Fuel reformers must also be factored into the efficiency of systems operating on logistics fuel (JP-8). The fuel utilization factor for the reformer is written as follows:

$$\mu_f = \text{LHV of fuel products out/LHV of fuel in}$$

Because all reformers operate above the boiling point of water, the LHV of the fuels is always used.

The fuel utilization, μ_f, in fuel cells and engines can be calculated from the ratio of the fuel converted to electricity by electrochemical oxidation to the fuel provided. For heat-producing engines, μ_f can be defined as the ratio of oxidized fuel producing heat (for potential conversion to work) to the total fuel content input to the engine.

In fuel cells, fuel is not fully utilized for several reasons. The greatest loss comes from incomplete electrochemical oxidation of the fuel at the anode. In closed systems, such as direct methanol fuel cells (DMFCs), where the fuel is recycled, the unreacted fuel can be passed back over the anode, but even then the fuel cell can never convert all of its fuel to electricity because some of it is vented with the exhaust from the anode (H_2O and CO_2) and there must be some extra fuel carried. Low-temperature fuel cells with polymer electrolytes (PEM fuel cells and DMFCs) also experience crossover of fuel from the anode through the electrolyte to the cathode, where the fuel is oxidized to heat and reaction products and no useful electrons (see Appendix D). In some designs, the fuel can also be oxidized by internal currents within the cell stack or by shunt currents, which effectively short circuit the external load and therefore produce no energy. Fuel can also be lost by mechanical means, such as purging or evaporation.

All of the factors above can have a significant impact on overall system efficiency. For a low-temperature PEM fuel cell system that produces liquid water, that is run at 0.8 V, and that has a 95 percent efficient BOP, the overall thermodynamic system efficiency can approach 49 percent. DMFCs are typically on the order of 30 percent efficient. They have lower overall efficiencies relative to PEM/H_2/air systems owing to a combination of lower cell voltages (0.5 V/1.25 V), reduced fuel utilization (μ_f <90 percent), and perhaps a higher BOP burden (η_{BOP} <90 percent). The efficiency parameters also vary with cell power or cell voltage, as shown in Figure C-3 for a nominally 20-W DMFC.

To simplify the characterization, fuel cell systems are also defined by their overall thermal efficiency, which is calculated from the following equation:

$$\eta_{el} = \text{Electrical energy output/Heating value of fuel in}$$

The efficiency values for fuel cells used in Chapter 2 are the overall thermal efficiency.

COMPARING PROSPECTIVE MILITARY SYSTEMS

Technology readiness levels (TRLs) are used by the Army and other government agencies to measure the relative maturity of system developments. Definitions for the nine levels used by the committee to compare prospective power systems are shown in Table C-1. TRL 6 is the level at which a power source model or prototype has been demonstrated in a relevant environment.

Once new power systems such as fuel cells and engines are developed, the efficiency of the overall system can be used to verify the validity of the system by predicting the maximum allowable weight of the hardware components, as described below.

First, the calculated weights are given in Tables C-2 and C-3 for 24- and 72-hr missions, respectively, for complete power sources operating at average levels of 2, 20, 100, and 3,000 W at specific energies of 200, 600, 1,000, 2,000, and 3,000 Wh/kg. For example, a 24-hr mission at 20 W has a net energy, E_{net}, of 480 Wh, and the power source would have a total weight of 2.4 kg (including system mass and fuel) if it has a specific energy of 200 Wh/kg and 0.48 kg if it has 1,000 Wh/kg. A 1,440-Wh mission, or 72 hr at 20 W, would require a 200-Wh/kg power source weighing 7.2 kg or a 1,000-Wh/kg power source weighing 1.44 kg. For fueled systems, such as fuel cells and engines, a portion of the total weight, W_{total}, is due to the system, W_{system} (including the fuel conversion system and associated balance of plant), and the remaining weight is due to fuel, W_{fuel}. W_{system} can be calculated from the system electrical efficiency, η_{el}; the net energy of the mission, E_{net}; the specific energy of the system, E_{sp}; and the heating value of the fuel, HV, by solving two equations:

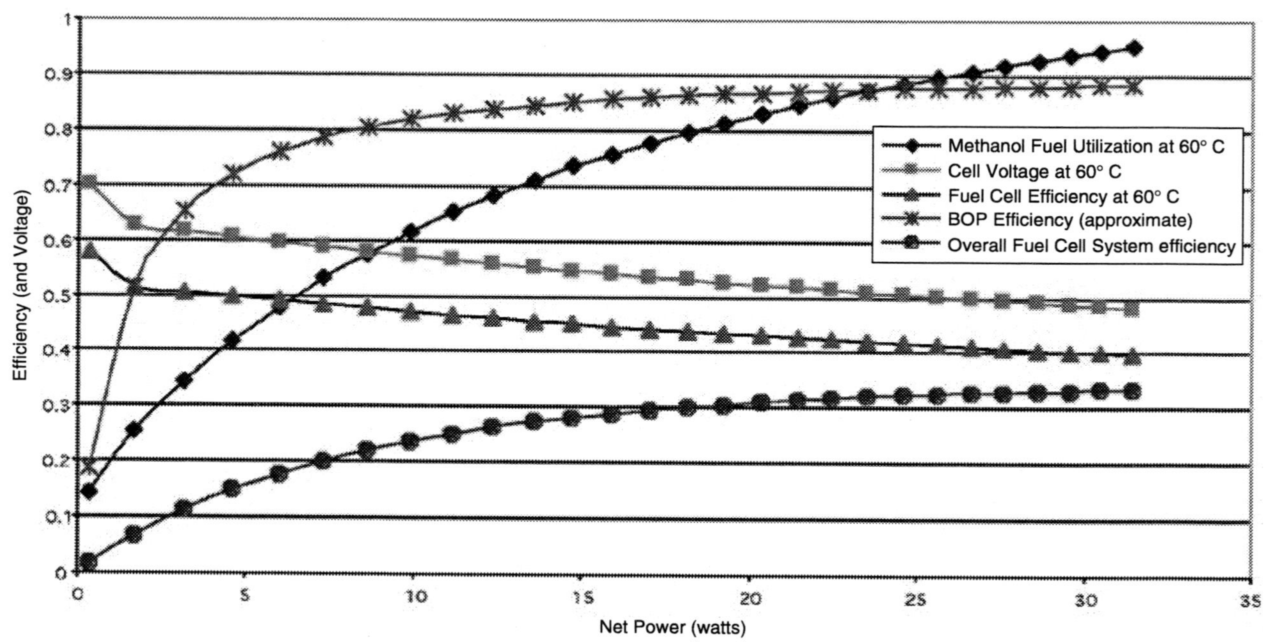

FIGURE C-3 Variation in efficiency parameters of a 20-W-rated DMFC with variations in the load (net power).

TABLE C-1 Criteria for Technology Readiness Levels

TRL	Task Accomplished	Description
1	Basic principals observed and reported	Lowest level of technology readiness. Scientific research begins to be translated into applied research and development. Examples might include paper studies of a technology's basic properties.
2	Technology concept or application formulated	Invention begins. Once basic principles are observed, practical applications can be invented. The application is speculative and there is no proof or detailed analysis to support the assumption. Examples are still limited to paper studies.
3	Analytical and experimental critical function or characteristics proof of concept	Active research and development are initiated. These include analytical studies and laboratory studies to physically validate analytical predictions of separate elements of the technology. Examples include components that are not yet integrated or representative.
4	Component or breadboard validation in laboratory environment	Basic technology components are integrated to establish that the pieces will work together. This is relatively "low fidelity" compared with the eventual system. Examples include integration of ad hoc hardware in a laboratory.
5	Component or breadboard validation in relevant environment	Fidelity of breadboard technology increases significantly. The basic technological components are integrated with reasonably realistic supporting elements so that the technology can be tested in a simulated environment. Examples include high-fidelity laboratory integration of components.
6	System/subsystem model or prototype demonstration in a relevant environment	Representative model or prototype system, which is well beyond the breadboard tested for TRL 5, is tested in a relevant environment. Represents a major step up in a technology's demonstrated readiness. Examples include testing a prototype in a high-fidelity laboratory environment or in a simulated operational environment.
7	System prototype demonstration in an operational environment	Prototype near or at planned operational system. Represents a major step up from TRL 6, requiring the demonstration of an actual system prototype in an operational environment, such as in an aircraft, vehicle or space. Examples include testing the prototype in a testbed aircraft.
8	Actual system completed and flight qualified through test and demonstration	Technology has been proven to work in its final form and under expected conditions. In almost all cases, this TR represents the end of true system development. Examples include developmental test and evaluation of the system in its intended weapon system to determine if it meets design specifications.
9	Actual system flight proven through successful mission operations	Actual application of the technology in its final form and under mission conditions, such as those encountered in operational test and evaluation. In almost all cases, this is the end of the last bug fixing aspects of true system development. Examples include using the system under operational mission conditions.

SOURCE: NRC, 2003.

$$E_{sp} = E_{net}/W_{total} = E_{net}/(W_{system} + W_{fuel})$$

$$W_{system} = E_{net}/E_{sp} - W_{fuel} \quad (1)$$

and

$$\eta_{el} \times HV = E_{net}/W_{fuel}$$

$$W_{fuel} = E_{net}/\eta_{el} \times HV \quad (2)$$

Substitute equation 2 into equation 1 to get the final equation:

$$W_{system} = (E_{net}/E_{sp}) - [E_{net}/(\eta_{el} \times HV)]$$

Alternatively, if the dry system mass is known, its specific energy can be calculated as a function of its efficiency using the equation

$$E_{sp} = E_{net} \times \eta_{el} \times HV/(W_{system} \times \eta_{el} \times HV + E_{net})$$

These equations can be used to calculate the maximum allowable dry weight of a system (i.e., weight of the system without fuel) or the maximum allowable specific energy of a system with a known dry mass. The results of the equations are shown graphically in Figures C-4 and C-5 for 24-hr and 72-hr 20-W missions (net energy = 1,440 Wh) on methanol (HHV = 6,088 Wh/kg) and JP-8 (LHV = 12,000 Wh/kg) for devices having specific energies from 200 to 3,000 Wh/kg.

TABLE C-2 Energy and Total System Weights for 24-Hour Missions

		Weight of Total System (kg)[a]				
Power of Mission (W)	Energy of 24-hr Mission (Wh)	200 Wh/kg Device	600 Wh/kg Device	1,000 Wh/kg Device	2,000 Wh/kg Device	3,000 Wh/kg Device
2	48	0.24	0.08	0.048	0.024	0.016
20	480	2.4	0.8	0.48	0.24	0.16
100	2,400	12	4	2.4	1.2	0.8
3,000	72,000	360	120	72	36	24

NOTE: Calculated weights of devices having specific energies of 200, 600, 1,000, 2,000, and 3,000 Wh/kg at power levels of 2, 20, 100, and 3,000 W.
[a]System weight includes dry system plus fuel.

TABLE C-3 Energy and Total System Weights for 72-Hour Missions

		Weight of Total System (kg)[a]				
Power of Mission (W)	Energy of 72 hr Mission (Wh)	200 Wh/kg Device	600 Wh/kg Device	1,000 Wh/kg Device	2,000 Wh/kg Device	3,000 Wh/kg Device
2	144	0.72	0.24	0.144	0.072	0.048
20	1,440	7.2	2.4	1.44	0.72	0.48
100	7,200	36	12	7.2	3.6	2.4
3,000	216,000	1,080	360	216	108	72

NOTE: Calculated weights of devices having specific energies of 200, 600, 1,000, 2,000, and 3,000 Wh/kg at power levels of 2, 20, 100, and 3,000 W.
[a]System weight includes dry system plus fuel.

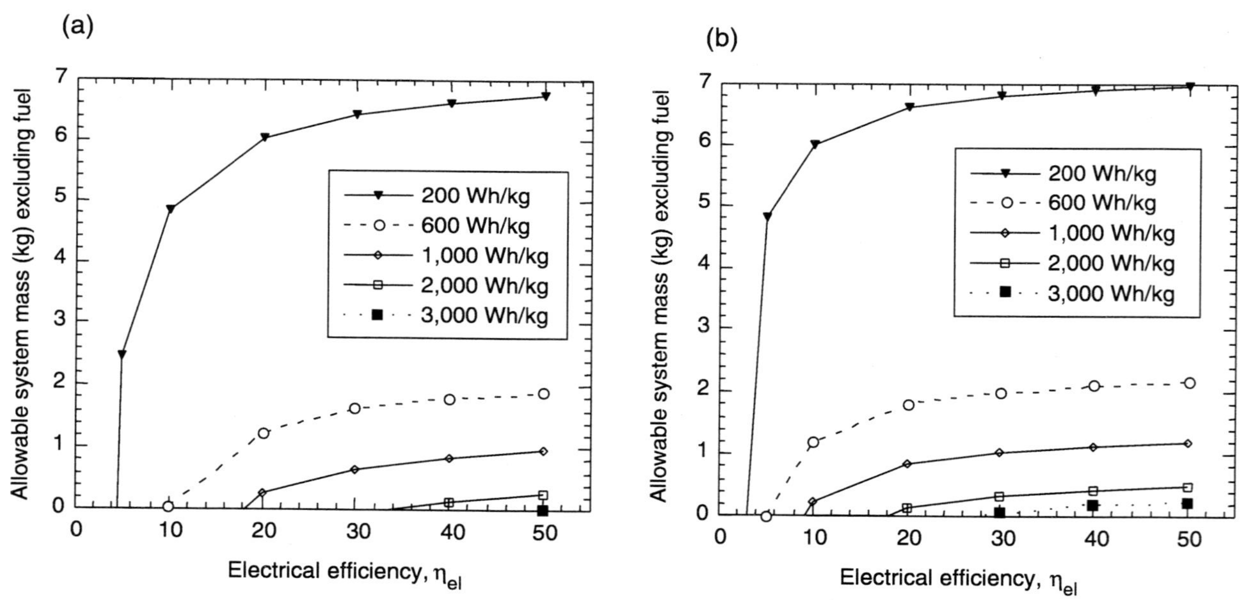

FIGURE C-4 The maximum allowable system mass (excluding fuel) calculated for (a) energy conversion systems operating below 100°C on methanol (HHV = 6,088 Wh/kg) or (b) high-temperature conversion systems (>100°C) operating on JP-8 (LHV = 12,000 Wh/kg). The energy density of the systems is varied from 200 to 3,000 Wh/kg. The data are derived from the equations in this appendix.

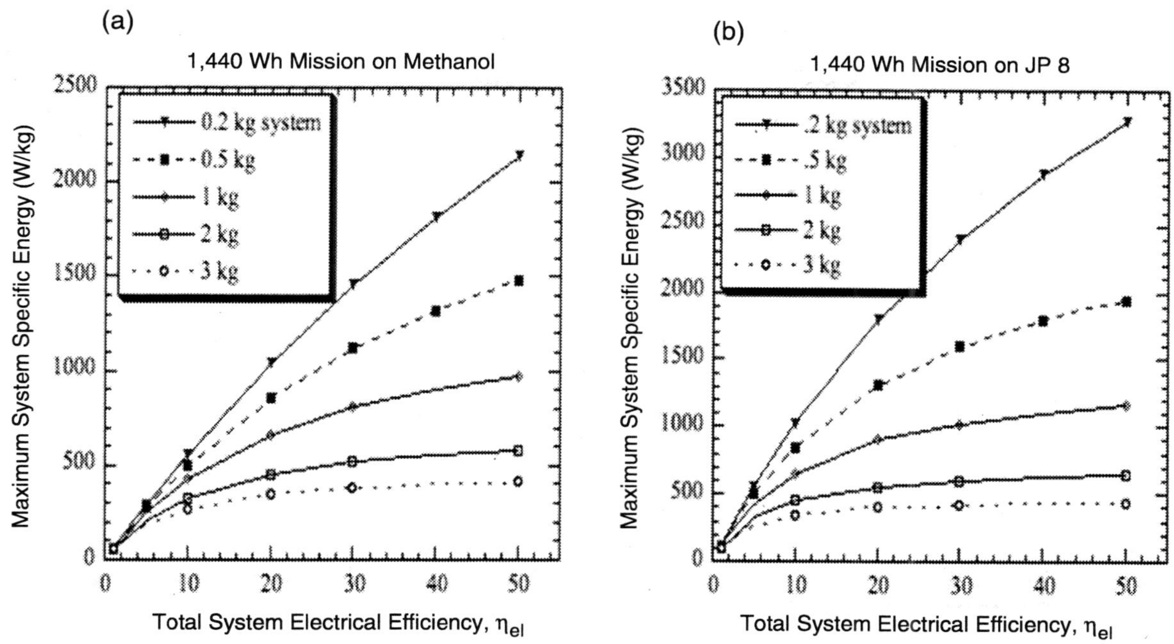

FIGURE C-5 The maximum allowable system specific energy calculated for (a) energy conversion systems operating below 100°C on methanol (HHV = 6,088 Wh/kg) or (b) high-temperature conversion systems (>100°C) operating on JP-8 (LHV = 12,000 Wh/kg). The dry weight of the systems is varied from 0.2 to 3 kg. The data are derived from the equations in this appendix.

The calculations for Figure C-5 are made with a simple equation, given below for a 2,000-Wh/kg system with a 1,440-Wh mission on methanol (6,088 Wh/kg):

$$\text{System mass (kg)} = (1{,}440/2{,}000) - [1{,}440)/(\text{efficiency}/100) \times 6{,}088]$$

A similar equation is used to generate the data plotted in Figure C-5 for a 2-kg dry system:

$$\text{System electrical efficiency (Wh/kg)} = \frac{1{,}440 \times (\text{efficiency}/100) \times 6{,}088}{[2 \times (\text{efficiency}/100) \times 6{,}088] + 1{,}440}$$

The plots in Figures C-4 and C-5 show that for a 20-W system used for 72 hr (1,440 Wh), a greater dry system mass can be tolerated by a fuel cell or engine that operates on JP-8 rather than methanol. A 30 percent efficient DMFC must weigh 1.6 kg to operate at 600 Wh/kg, but a SOFC running on JP-8 can weigh 2 kg. This discrepancy is simply due to the energy content of the fuels. One can infer from the plots that system efficiency and system weight are the key factors affecting the specific energy of the system. Also, systems with efficiencies below 10 percent are limited in their specific energy—for instance, a 1,000-Wh/kg system can never be achieved by a methanol-fueled system with 10 percent efficiency (Figure C-5)

The equations can be used to estimate the viability of certain systems for various missions. For a 480-Wh mission (20 W for 24 hr), any JP-8 fueled system with a target specific energy of 1,000 Wh/kg would have to weigh less than 400 g; the task of integrating the device, the insulation, the fuel tank, and so forth would be a challenging one. However, when the mission duration is extended to 4,800 Wh (10 days at 20 W), the fueled systems become highly attractive.

REFERENCES

Cairns, Elton J. 2004. Battery overview. Chapter B. Encyclopedia of Energy, Vol. 1. C.J. Cleveland, ed. New York: Elsevier, 124.

Chase, M.W. 1986. JANAF Thermochemical Tables, 3rd ed., vols. 1 and 2. American Chemical Society.

NRC (National Research Council). 2003. Science and Technology for Army Homeland Security: Report 1. Washington, D.C.: The National Academies Press.

Panasonic. 2003. Panasonic Nickel Metal Hydride Handbook. Available at http://www.panasonic.com/industrial/battery/oem/chem/nicmet/index.html. Last accessed on December 23, 2003.

Appendix D

Source Technologies

This appendix provides supporting information for the discussions of technology alternatives in Chapter 2, the hybrid systems in Chapter 3, and the advanced concepts in Chapter 6. It describes background information on new and advanced power sources in the 2-W, 20-W, 100-W, and 1- to 5-kW regimes that was not included in *Energy-Efficient Technologies* (NRC, 1997). Table D-1 provides a comprehensive list of the source technologies discussed in both reports.

BATTERIES

Batteries are electrochemical devices that convert the chemical energy of active materials into electrical energy. A battery cell comprises a negative electrode (anode) and a positive electrode (cathode) having differing electrical potentials; these electrodes are electronically separated but are ionically connected with an electrolyte. Current collectors, packaging, and interconnects are needed to deliver the energy safely to a load. This ensemble is shown schematically in Figure D-1. The arrangement or geometry of the cell has a significant impact on the discharge properties of the cell. An excellent overview of battery chemistries, their definitions, design, and properties is available at http://voltaicpower.com.[1] Most battery manufacturers also have detailed descriptions of their batteries' chemistries and properties. Also see sources such as the *Handbook of Batteries* (Linden and Reddy, 2002).

Primary Batteries

Primary batteries can be discharged once and then must be discarded. Most primary battery technologies are very mature, but there are several systems that might be improved to the point where they could have a significant impact on the military. The R&D efforts for commercial batteries are concentrated on the design of new form factors for specific device applications and on the search for materials capable of high-energy/high-power performance. Research on less mature chemistries can still yield improvements in power and energy.

The properties of commercial Li/SO_2, Li/MnO_2, and $Li/(CF)_x$ batteries are summarized in Table D-2. The military uses Li/SO_2 batteries for many applications, specifically in the BA 5590, which is the workhorse of soldier electronics. These have a theoretical voltage of 3.1, a working voltage of 2.8, and a practical energy density of 170 Wh/kg. D-cell configurations on Li/SO_2 batteries have specific energies of 210 Wh/kg (e.g., SAFT LO26SX). The trend is to replace the Li/SO_2 batteries with Li/MnO_2 batteries, which have fewer safety constraints.

The Li/MnO_2 battery is a commercially available primary system, and Li/MnO_2 button cells (123A and 223A) are used for small device applications such as watches, calculators, cameras, and clocks. The theoretical voltage of the reaction is 3.5, but it has a practical voltage of 3.3. Its electrolyte is an organic solvent with a Li salt. The military is currently taking orders on Li/MnO_2 batteries for use in SINCGARS radios and the like from SAFT[2] and Brentronics (BA-5372/U, 5368/U, X567/U). The properties of commercial Li/MnO_2 batteries are summarized in Table D-2.

The $Li/(CF)_x$ cell was first introduced in Japan by Matsushita (Panasonic[3]) in the early 1970s. $Li/(CF)_x$ coin cells and BR 2/3A cells are two popular commercial cells. $Li/(CF)_x$ coin cells are used mainly in low-drain devices such as electronic watches and calculators. BR 2/3A cells are used

[1]Last accessed on January 28, 2004.

[2]Found at www.saftbatteries.com. Last accessed on January 28, 2004.

[3]Found at http://www.panasonic.com/industrial/battery/oem/chem/lithion/index.html. Last accessed on January 28, 2004.

TABLE D-1 Overview of All Power Source Alternatives

Power System	State of the Art, 1997[a]	State of the Art, 2003	Item Considered	Scaling Laws	Impact on Soldier Power
Primary battery (includes metal/air)	Mature. Up to 800 Wh/kg in low-specific-power configurations	Mature. SOA not significantly advanced beyond NRC (1997) report.	Energy density. Safety. Power density. Environmental impact.	Known	Heavy, one-time use. Current battery of choice for combat missions. Potential for use in hybrids.
Secondary battery	Mature. Li ion: 100 Wh/kg in development.	Mature in commercial applications. Li ion: 140 Wh/kg available; 200 Wh/kg in development.	Energy density. Cycle life. Power density. Safety and cost.	Known	Stand-alone energy supply for many missions. Can be used in hybrid mode for high-energy missions.
Fuel cell (hydrogen)	Exploratory development. Many systems at laboratory scale. Power levels to 150 W considered.	Beta prototypes with various hydrogen sources tested in field. Power to 150 W.	Fuel reformers. Water management. Safety.	Known	New capability; potential for use in hybrid system. Less weight. Cost savings. Requires new battlefield fuel.
Fuel cell (methanol)	Emerging. Not considered.	Beta prototypes developed at power levels of 20 to 50 W. 20% efficiency.	Fuel and fuel crossover. Catalyst. Cost.	Known	New capability. Less weight. Cost savings. Requires new battlefield fuel.
Fuel cell (solid oxide)	Emerging. Not considered.	Emphasis on small sizes. Laboratory prototypes in 20-W range. Research in high-capacity designs.	High temperature. Materials. Integration and systems.	Known	New capability. Less weight. Easier to utilize battlefield fuels. More efficient.
Internal combustion	Some versions mature. Hobby application sizes coupled to generators. No commercial products on market.	Commercial applications with motor-alternator combinations in 30 to 100 W/kg range. Efficiencies greater than 20% in 500-W sizes. Emerging modified hobby engines operate on diesel.	Fuels. Vibrations. Life.	Known	Inexpensive technology. Potential for high-energy missions. Can probably be made to function with JP fuels. Current role as battery charger.
External combustion (includes Stirling)	Not considered.	100 W/kg specific power demonstrated for motor-alternator with efficiency of 29%. System efficiencies projected to be >20%. Laboratory 35- to 50-W systems available for beta prototypes; 1- to 2-kW beta prototypes available with ~20% system efficiencies. System-specific power appears to be around 30 W/kg.	Fuels. Specific power. System-specific energy. Signatures.	Known	New stealth capability. Inexpensive technology. Can be made to operate on JP fuels. Potential for high-energy missions.
Microturbine	Emerging. Considered promising.	Not considered owing to lack of progress in producing workable systems.	Fuels. Specific power. System-specific energy. Materials. Cost.	Unknown	

continued

TABLE D-1 Continued

Power System	State of the Art, 1997[a]	State of the Art, 2003	Item Considered	Scaling Laws	Impact on Soldier Power
Thermoelectric	Some versions mature. Low potential. Best system efficiency on order of 5%; converter efficiencies projected to 10%.	Insufficient progress to consider for current applications. Progress in new high-ZT materials makes technology worth watching for long term.	Efficiency. Materials-specific power. System-specific energy.	Known	Not applicable owing to low efficiency. Possible niche application in small sizes.
Thermo-photovoltaic (TPV)	20% TPV cells demonstrated. System projections to 20%.	Not considered owing to lack of progress in systems.		Known	
Nuclear isotope	Limited consideration. Rejected owing to cost, safety, environmental considerations, and lack of infrastructure.	Not considered.	Safety. Environmental impact. Cost. Public acceptance.	Known	
Alkali metal thermal-to-electric converter	Speculative technology. Systems projection to 500 W/kg.	Not considered owing to lack of progress.		Known	
Energy harvesting; solar	Some versions mature.	Considered for low-capacity niche applications.		Known	Driver for reducing power demand.

NOTE: SOA, state of the art; Li ion, lithium ion; JP, jet propellant; ZT, thermoelectric figure of merit.
[a]NRC, 1997.

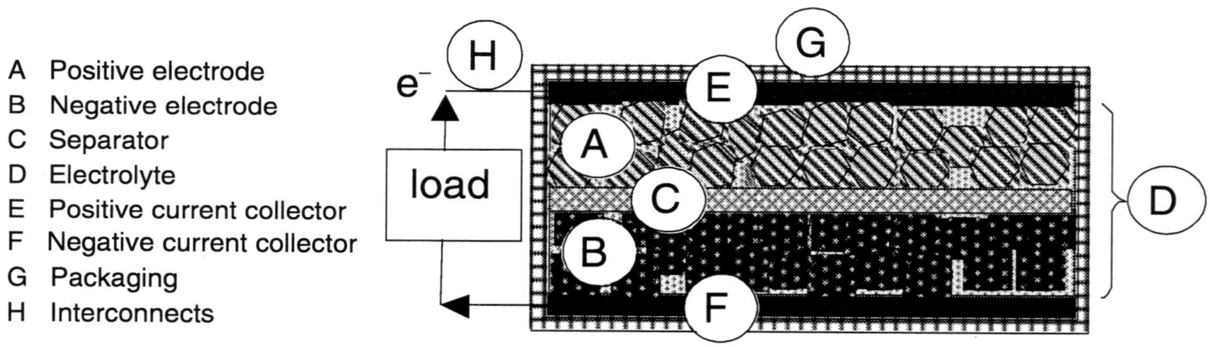

FIGURE D-1 Schematic cross section of a battery.

in cameras, though the Li/MnO_2 2/3A cell is the more popular choice for cameras due to its lower cost. As summarized in Table D-2, $Li/(CF)_x$ has higher theoretical specific energy than Li/MnO_2 cells (2120 Wh/kg vs. 900 Wh/kg) and an open circuit potential (OCV) of about 3.2 V. The theoretical OCV, based on free-energy calculations, is about 4.5 V. The difference between theoretical and practical OCV values has been discussed by Whittingham (1975). A comparison of the practical performance of $Li/(CF)_x$ vs. that of Li/MnO_2 is shown in Table D-2.

In spite of the much higher theoretical specific energy in a $Li/(CF)_x$ cell, $(CF)_x$ is much lighter than MnO_2 (2.5 g/cc vs. 4.5 g/cc) and gives comparable practical energy performance in commercial small cells. During the discharge of the cell, the carbon monofluoride in the positive electrode changes from a poor conductor to a more conductive amorphous carbon when discharged. Thus, the reaction efficiency increases with discharge. $Li/(CF)_x$ cells are known for their high-temperature performance (as high as 150°C according to Panasonic coin cells data), long shelf life (>10 years), and

TABLE D-2 Attributes of Advanced Primary Batteries

Attribute	Chemistry: Lithium Sulfur Dioxide (Li/SO_2)	Lithium Manganese Dioxide (Li/MnO_2)	Lithium Carbon Monofluoride ($Li/(CF)_x$)
Discharge reaction	$2Li + 2SO_2 \rightarrow Li_2S_2O_4$	$xLi + Mn^{IV}O_2 \rightarrow Li_xMn^{III}O_2$	$xLi + (CF)_x \rightarrow xLiF + C$
Theoretical voltage (V)	3.10	3.50	4.50
Working voltage (V)	2.95	3.30	3.50
Energy density (Wh/L)[a]	385	480-510	1,040
Specific energy (Wh/kg)[a]	210	210-250	600
Power density (W/L)	<180	<230	<23
Specific power (W/kg)	<100	<100	<14
Shelf life		5 yr	>10 yr
Reference	SAFT LI26SX	Duracell 2/3A	Eagle-Picher LCF-112
Cell capacity (Ah)	7.5	1.4	39.4

[a]The energy density and specific energy values are based on density and specific power values, respectively.

TABLE D-3 Attributes of Leading Secondary Batteries

Attribute	Chemistry: Lithium Ion	Nickel Metal Hydride (MH/NiOOH)	Lithium/Sulfur
Negative electric discharge	$LiC_6 = Li^+ + C_6 + e^-$	$MH + OH^- = M + H_2O + e^-$	$Li = Li^+ + e^-$
Positive electric discharge	$Li\frac{1}{2}CoO_2 + \frac{1}{2}Li^+ + \frac{1}{2}e^- = LiCoO_2$	$NiOOH + H_2O + e^- = Ni(OH)_2 + OH^-$	$S_x + 2e^- = S_x^=$
Overall reaction	$LiC_6 + 2Li\frac{1}{2}CoO_2 = C_6 + 2LiCoO_2$	$MH + NiOOH = Ni(OH)_2 + M$	$2Li + S_x = Li_2S_x$
Theoretical voltage (V)	~4.2	1.2	2.1
Working voltage (V)	3.6	1.0	1.8
Cost (initial, $/Wh)	~10	~3	~0.25
Energy density (Wh/L)[a]	450-490	220	225
Specific energy (Wh/kg)[a]	160-175	63-75	170
Power density (W/L)	<570	850	50
Specific power (W/kg)	<200	220	50
Life cycles	300-1,000	600-12,000	300-650
Environment (°C)	–20 to +60°C	–30 to +65°C	+25 to +60°C
Reference	Sanyo 18650	Linden and Reddy (2002)	Polyplus 1 Ah cells

[a]The energy density and specific energy values are based on the power density and specific power values, respectively.

high specific energy at low to medium powers. In comparison with Li/MnO_2, the main disadvantages of $Li/(CF)_x$ are low power capability and high cost.

Secondary Batteries

Secondary batteries can be recharged. There are numerous commercially available secondary batteries that are used commercially, such as lead-acid, silver-zinc, and metal-hydride systems. This appendix describes systems that have advanced technologically since 1997, including Li ion and Li polymer chemistries, nickel metal hydride, and lithium sulfur. Attributes of these batteries are summarized in Table D-3.

Li ion batteries encompass several different chemistries, including $LiCoO_2$, $LiNiO_2$, and $LiMn_2O_4$ positive electrodes. The Li ion cell was introduced commercially in the early 1990s by the Sony Corporation.[4] It has the advantages of high cell voltage (~3.6 V), high specific energy (>100 Wh/kg), and long cycle life (~1,000 deep cycles). Li ion batteries' power and energy characteristics are summarized in Table D-3. Li ion batteries quickly captured the market for camcorders, cell phones, and notebook computers in spite of their high cost, and small cells of cylindrical and prismatic form are being manufactured at the rate of close to a billion cells per year.

The cells can be recharged because the active materials can accommodate the movement of Li atoms (and electrons)

[4]Found at http://www.sanyo.com/industrial/batteries/. Last accessed on January 28, 2004.

into and out of the structure, with a minimum of disruption to that structure. This structural integrity is important in maintaining a long cycle life. The negative electrode is made of various types of carbon and graphite (the original Sony cell used $LiCoO_2$). The CoO_2 has a layered structure that readily accommodates the Li without the formation of a new structure (or new phase).

Although Li ion cells have the best performance of any available rechargeable battery, they have a number of problems that are currently being addressed by the R&D community. Overcharge or overdischarge can lead to capacity loss and even cell failure in the form of thermal runaway and fire, so each cell has a protective microcircuit that controls the voltage limits of the cell and the recharge process. The solvents for the electrolyte are flammable organic liquids (such as ethylene carbonate and dimethyl carbonate), so there is research on flame-retardant additives. Also, because the cobalt oxide positive electrodes are expensive, alternative low-cost, high-capacity positive electrode materials are being explored, including $LiNiO_2$-based, $LiMnO_2$-based, $Li(Mn,Ni,Co)O_2$-based, and $LiFePO_4$-based materials. Some nickel-containing materials are close to commercialization. Performance can also degrade by spontaneous film formation on the electrodes, so there are efforts to find additives for the electrolyte that control film formation and film properties.

Li polymer cells are derivatives of the Li ion cells. They have the same electrochemistry, but the liquid electrolyte is gelled with a polymer such as polyvinylidene fluoride (PVdF) or polyethylene oxide so that it is immobilized and behaves like a polymer. The gel offers flexibility in the shape of the cell and eliminates any free-flowing liquid. Li polymer cells have performance similar to that of the Li ion cell, with specific energy values up to about 150 Wh/kg and 300 Wh/L for –20°C to +60°C, and have been recently introduced to the commercial market.[5]

The nickel metal hydride, or MH/NiOOH, cell has become very popular for many consumer applications, including portable electronics and power tools. It has largely replaced the Ni-Cd (Cd/NiOOH) cell in the consumer market, because of concern about the environmental impact of cadmium. The MH/NiOOH cell has an aqueous electrolyte of potassium hydroxide, which offers a much higher conductivity than the nonaqueous electrolytes used in lithium cells, so it can be discharged at high power. Both of the electrode reactions in Table D-3 are reversible and have rapid reaction rates, so high specific power values can be achieved, but their specific energy is less than 100 Wh/kg, which limits its usefulness. Other problems with this system include its low cell voltage (~1.2 V), limited temperature range for reasonable operation, and the need for charging at a relatively low temperature (<45°C).

[5]Found online at http://www.ulbi.com/product-grid.asp. Last accessed on January 28, 2004.

Li/S cells offer the opportunity for very high specific energy (theoretical value = 2,600 Wh/kg) and low cost, using environmentally benign materials. Their characteristics are summarized in Table D-3. The drawback of this battery system is its short cycle-life, which is due to the sulfur electrode. During operation of the cell, polysulfides of several stoichiometries form and dissolve in liquid electrolytes, allowing them to migrate throughout the cell. This stability issue has been addressed by using gel and polymer electrolytes that prevent migration of the sulfur species. Sion Power Corporation[6] is striving to introduce commercial lithium/sulfur batteries in 2004 with 1-Ah pouch-style cells.

Metal/Air Batteries

Metal/air cells comprise a cathode that uses oxygen in the air as an oxidant and a solid fuel as the anode. They are different from fuel cells and other batteries in that the anode is consumed during operation. Often, metal/air cells are described as semi-fuel cells. Metal/air cells are being studied because they have the advantage of using air as an inexhaustible cathode reactant, leading to compact, anode-limited cells with high energy density. Carbon/air batteries are grouped with this class of power sources even though they operate at elevated temperatures.

The properties of metal/air and carbon/air electrochemical couples are summarized in Table D-4. The total metal/air reaction is the sum of the reaction of the oxidation at the metal anode and the reduction of oxygen at the air cathode:

$$4M + nO_2 + 2nH_2O \rightarrow 4M(OH)_n$$

$$M + nO_2 \rightarrow MO_{2n}$$

where M is the metal and n depends on the valence change for the oxidation of the metal. Most metal/air cells do not have a long shelf life once they are activated with electrolyte and exposed to air, because the metal anode tends to react with water in the aqueous electrolyte or moisture in the air to generate hydrogen:

$$M + nH_2O \rightarrow M(OH)_n + n/2\ H_2$$

Moisture in the air is a big factor in the performance of metal/air cells. Too much moisture causes flooding of the air electrodes, while insufficient moisture causes water to evaporate from the cells and dries out the electrolyte. In addition, metal/air cells that use alkaline electrolyte also suffer from the buildup of carbonates in the electrolyte from the reaction with CO_2 in the air. Finally, the slow gas-solid

[6]Found online at http://www.sionpower.com. Last accessed on January 28, 2004.

TABLE D-4 Attributes of Metal/Air and Carbon/Air Batteries

Attribute	Lithium/Air	Aluminum/Air	Magnesium/Air	Zinc/Air	Carbon/Air
Discharge reaction	$2\,Li + H_2O + \frac{1}{2}O_2 \rightarrow 2LiOH$	$4Al + 3O_2 + 6H_2O \rightarrow 4Al(OH)_3$	$Mg + \frac{1}{2}O_2 \rightarrow MgO$	$Zn + \frac{1}{2}O_2 \rightarrow ZnO$	$C + O_2 = CO_2$
Theoretical voltage (V)	3.40	2.70	3.10	1.60	1.00
Working voltage (V)	2.85	1.10-1.40	1.60	1.00-1.20	
Theoretical specific energy of metal/fuel (Wh/kg)	13,000	8,100	6,800	1,300	9,100
Specific energy (Wh/kg)	2,600 (est.)	1,620 (est.)	700	260	2,400 (projected)

diffusion of oxygen at the cathode makes most metal/air cells suitable only for low-to-moderate specific energy sources.

There is renewed interest by the Army in studying the feasibility of Li/air cells due to their theoretically high energy density and specific energy. The Li/air chemistry is attractive because it combines Li, the electronegative material with the highest capacity, with air. Li has a theoretical specific energy of 13,000 Wh/kg assuming a theoretical cell voltage of 3.4 V, though only 2.85 V is achieved in practice. The Li/air reaction is given in Table D-4. Due to the high reactivity of Li with water, the undesirable competing reaction is

$$Li + H_2O \rightarrow LiOH + {}^1/_2\,H_2$$

Besides the undesirable high reactivity of Li with water, the kinetics of oxygen diffusion through the cathode also limits the Li/air cells, although recent efforts in Li/oxygen rechargeable cells and fuel cells should improve the kinetics of the air cathode in Li/air cells. Following the pioneering work of the EIC group on developing the Li/O_2 battery, Read et al. (2003) found that the oxygen solubility in the electrolyte and the electrolyte viscosity had direct impact on the discharge rate of Li/O_2 cells (Abraham and Jiang, 1996; Read et al., 2003).

Promising results on Li/air cells were obtained recently at PolyPlus Battery Company.[7] Using a novel protective coating on the Li, researchers at PolyPlus were able to demonstrate complete discharge of a 50-μm-thick Li anode at 0.3 mA/cm^2 in air. This preliminary result suggests that the corrosion and rate issues with the Li/air system can be resolved, but further research is needed to evaluate the feasibility of scaling up this technology and the stability of the coated Li in extended storage.

Aluminum also has a high specific energy, 8,100 Wh/kg, assuming a theoretical cell voltage of 2.7, though the voltage of Al/air cells is about 1.3 in practice. As with Li/air cells, Al reacts with water in the electrolyte to form $Al(OH)_3$ and hydrogen gas. In practice, Al/air cells can use either neutral (saline) or alkaline electrolyte. The saline electrolytes have low corrosion rates and are used mainly for low-power applications. Al/air cells with alkaline electrolytes are high-rate cells owing to the high conductivity of the electrolyte, but they also exhibit high corrosion rates. Thus, alkaline Al/air batteries are often used as reserve batteries that are activated before use by adding the electrolyte. For portable military applications, saline Al/air cells might be useful in a hybrid configuration as an energy source. Work is needed to develop Al alloys that are less reactive with water and to develop electrolyte formulations in which $Al(OH)_3$ is less soluble in order to minimize loss of electrolyte conductivity.

Magnesium has a specific energy of 6,800 Wh/kg, assuming a theoretical cell voltage of 3.1 in Mg/air cells, although the actual cell voltage is about 1.6 in practice. In alkaline electrolyte, the Mg is passivated by the formation of $Mg(OH)_2$. The insoluble surface film of $Mg(OH)_2$ protects the Mg from further reaction with water but causes a voltage delay, seen also in $Li/SOCl_2$ batteries. Mg/air batteries were not commercialized in the past and were used mostly for undersea, low-rate applications, with 700 Wh/kg demonstrated. Mg/air cells were designed to deliver 3-4 W for one year or longer. Recently, attempts were made by Evionyx to commercialize Mg/air cells.

Zn/air batteries are being considered by the Army for hybrid systems; they are commercially available in button format for use in hearing aids. The theoretical specific energy is 1,300 Wh/kg, but in an operational cell 260 Wh/kg can be expected. The Zn/air system is subject to capacity loss due to leakage, electrolyte dry out, and carbonation, problems that have never been solved sufficiently for the battery to have a long life once it is activated. Those problems have not prevented Electric Fuel from making the 30/60 Ah BA-8180/U Zn/air battery for the Army.

Carbon/air batteries are in theory attractive, because carbon is an energy-dense fuel with a specific energy of 9,100 Wh/kg, and the batteries are safe to carry and nontoxic. Researchers have tried for decades to design devices for the electrochemical conversion of carbon to electricity, and progress has been made toward this goal in recent years. These systems are similar to other metal/air systems in that

[7]Found online at http://www.polyplus.com/. Last accessed on January 28, 2004.

they use a solid fuel and air as the oxidizer, but they operate at elevated temperatures (>650°C) to fully oxidize carbon to carbon dioxide. Because the final reaction product, CO_2, is a gas, these systems do not suffer over time from the buildup of solid reactant products, as is the case with Al, Li, Mg, and Zn systems. The efficiency of the energy conversion process is calculated to be in excess of 80 percent due to the lower heating value (LHV) of carbon. The operating temperature and cell efficiency are a function of the activity of the cell electrodes and of the type of electrolyte, with most designs utilizing electrolytes of either molten carbonate or solid oxide.

Lawrence Livermore National Laboratory (LLNL) has been studying carbon/air batteries and fuel cells as power sources. LLNL has made innovations in its anode and carbon fuel to achieve cells with significant power densities of up to 500 mA/cm^2 at 0.8 V, and they project energy values for their system in excess of 2,400 Wh/kg and 900 Wh/L. Because of these preliminary data, carbon-air batteries were identified as a top technology at the ARL/CECOM Energy and Power Workshop of October 2002. However, progress on carbon/air batteries is still at an early stage, and no system has ever been fully designed and integrated, even at the breadboard stage. Key challenges remain—for instance, in the thermal management of the cells, the methods to continuously feed the carbon to the cell anode, and the start-up time of the cells. It is too early (TRL = 2) to accurately predict the contribution of carbon/air systems to the Army, although basic research in this area is worthwhile.

ELECTROCHEMICAL CAPACITORS

There has been a surge of interest in electrochemical capacitors (supercapacitors or ultracapacitors, abbreviated EC), which produce one or two orders of magnitude more energy than traditional electrostatic capacitors. They are of particular interest for use in hybrid systems like those described in Chapter 3. In an electrostatic capacitor, the electrical energy is derived via charge accumulation and stored on the positive and negative plates, separated by a vacuum or a dielectric layer in a nonfaradaic process. In contrast, in a battery the electrical energy is derived from a change in the oxidation state of the active materials and is often accompanied by chemical changes to the structure via a faradaic process. The faradaic process is slow because it involves diffusion of ions into the bulk of active materials. Consequently, batteries usually are operated at lower power than ECs. However, ECs usually have less energy than batteries because most of the charge is stored near the surface layers of the electrodes and not in the bulk of the material. Finally, capacitors usually have a much longer cycle life than rechargeable batteries since the cycling process does not induce chemical or structural changes in the electrode materials. The failure of capacitors usually can be attributed to the breakdown of the dielectric layer or the electrolyte. The characteristics of ECs and batteries are compared in Table D-5.

ECs have electrolytes separating the two electrodes instead of the vacuum or dielectric layer present in electrostatic capacitors. The electrolyte not only serves as an ionic conductor but is also the source of ion separation and accumulation at the electrode/electrolyte interface. The electrolyte can be either aqueous or nonaqueous, but because nonaqueous electrolytes can be used with higher operating voltages, they lead to higher energy densities than aqueous electrolytes. Electrochemical capacitors can also be subdivided into asymmetric and symmetric types. In symmetric ECs, energy storage is nonfaradaic in both electrodes, but in asymmetric ECs charge storage in one of the electrodes is faradaic (or like a battery).

The maximum specific energy and power density for various types of capacitors are listed in Table D-6. Recent advances entailed the use of nano-materials such as nano-$Li_4Ti_5O_{12}$ to increase the power density of electrochemical capacitors (Amatucci, 2001). The energy density can be further increased by creating a battery + EC hybrid system with a mixture of activated carbon and a lithiated oxide (e.g., $LiCoO_2$) for the positive electrode and nano-$Li_4Ti_5O_{12}$ for the negative electrode (Amatucci, 2003). An energy density in excess of 30 Wh/kg at a power density of 3,000 W/kg can be obtained with such configuration. Such hybrid configurations help to bridge the gap between the energy and power characteristics of batteries and capacitors.

FUEL CELLS

Fuel cells are currently under intense research and development as power sources for a range of applications, including portable power, automobiles, and large-scale power plants. A fuel cell produces electrons via the electrocatalytic reduction and oxidation of an oxidizer and a fuel, respectively.[8]

For portable power sources, the proton exchange membrane fuel cell (PEMFC), the direct methanol fuel cell (DMFC), and the solid oxide fuel cell (SOFC) are the most attractive. The attributes of these three fuel cell systems (their operating temperatures, electrode reactions, and pros and cons) are given in Table D-7. The electrode reactions and operation of a PEMFC cell are shown schematically in Figure D-2: Hydrogen fuel is oxidized at the anode to protons that flow through a solid polymer electrolyte, and the protons

[8]Numerous texts are dedicated to fuel cells. An excellent basic resource is James Larminie and Andrew Dicks's Fuel Cell Systems Explained (John Wiley & Sons, Ltd, 2000). Web-based resources include http://www.fuelcells.org/fchandbook.pdf (last accessed on January 28, 2004), http://www.eere.energy.gov/hydrogenandfuelcells/education.html (last accessed on January 28, 2004) and http://voltaicpower.com/FuelCell/Frames.htm (last accessed on January 28, 2004).

TABLE D-5 Overall Comparison of Electrochemical Capacitor and Battery Characteristics

Capacitor Characteristics	Battery Characteristics
Intrinsically sloping charge and discharge curve.	Ideally, constant (thermodynamic) discharge or recharge potential, except for Li intercalation systems.
Because of preceding characteristic, has good intrinsic stage-of-charge indication.	Does not have good intrinsic state-of-charge indication except for Li intercalation systems.
Relatively poor energy density.	Moderate or good energy density, depending on equivalent weights and electrode potentials of active materials.
Good power density.	Relatively poorer power density, depending on kinetics.
Excellent cyclability or cycle life due to simple addition or withdrawal of charges (in double-layer type).	Less cycle life by a factor of 1/100 to 1/1,000 due to irreversibility of redox and phase-change processes in three dimensions.
Internal infrared (IR) due to high-area matrix and electrolyte.	Internal IR due to electrolyte and active materials.
Little or no polarization, but capacitor may be temperature-dependent.	Significant temperature-dependent activation polarization (faradaic resistance).
Long lifetime except for corrosion of current collectors and so on.	Poorer lifetime due to degradation or reconstruction of active materials.
Electrolyte conductivity can diminish on charging due to ion adsorption.	Electrolyte conductivity can decrease or increase on charging, depending on chemistry of cell reactions (e.g., with lead-acid).
Can be constructed in bipolar configuration.	Can be constructed in bipolar configuration.

SOURCE: Conway, 1999.

TABLE D-6 Attributes of Electrochemical Capacitors

Capacitor Type	Operating Voltage (V)	Maximum Specific Energy (Wh/kg)	Maximum Power Density (W/kg)	Cycle Life	Examples
Electrostatic	Frequency-dependent	0.01-0.05	10^7	$>10^6$	Mica, Mylar, paper
Electrolytic	Frequency-dependent	0.05-0.10	10^6	$>10^6$	Ta_2O_5, Al_2O_3
Symmetric electrochemical capacitor (aqueous)	0.9-1.2	7.16[a]	10^4	$>10^5$	Carbon/carbon
Symmetric electrochemical capacitor (nonaqueous)	2.0-2.5	9.41[a]	10^4	$>10^5$	Carbon/carbon
Asymmetric electrochemical capacitor (aqueous)	1.3-1.7	50.35[a]	10^4	$>10^5$	$Ni(OH)_2$/carbon
Asymmetric electrochemical capacitor (nonaqueous)	2.5-3.0	34.51[a]	10^4	$>10^5$	Carbon/$Li_4Ti_5O_{12}$

[a]Calculated data from Zheng, 2003.

recombine at the cathode via the reduction of oxygen to form water. Because the electrolyte only conducts ions, the electrons are forced through an external circuit and bear the potential of the voltage difference between the electrocatalytic reactions at the cathode and anode, minus ohmic losses. The electrolyte/electrode ensembles are referred to as membrane electrode assemblies (MEAs), with the fabrication of these MEAs having a significant bearing on their efficiency. Each fuel cell operates nominally between 0.5 and 0.9 V, and the system voltage is increased by stacking multiple cells together.

The principle advantages of fuel cells over other energy converter technologies (e.g., internal combustion engines) are the promise of fewer moving parts, longer life expectancy with less maintenance, lower operating pressures and temperatures, elimination of noxious emissions, and higher overall thermodynamic conversion efficiencies of fuel to electricity. The by-product of fuel cells is water, so they will

TABLE D-7 Attributes of Fuel Cells for Portable Power

Fuel Cell Type	Operating Temperature (°C)	Anode (Fuel) Reaction	Cathode (Oxygen) Reaction	Pros	Cons
Proton exchange membrane fuel cell	60-80	$H_2 = 2H^+ + 2e^-$	$\frac{1}{2} O_2 + 2H^+ + 2e^- = H_2O$	Prototype and commercially available units in a range of sizes (10 W to 1 MW). High power density. Amenable to rapid manufacturing. Rugged. High efficiency. Greatest government and commercial investment.	Hydrogen storage. Sensitive to poisoning. High cost. Difficult to operate from logistics fuels.
Direct methanol fuel cell	40-60	$CH_3OH + H_2O = CO_2 + 6H^+ + 6e^-$	$6H^+ + 6e^- + \frac{3}{2}O_2 = 3H_2O$	Prototypes available. Liquid fuel with no reformer. Significant government and commercial investment.	High cost. Low efficiency due to materials problems with catalysts and membrane.
Solid oxide fuel cell	700-1,000	On hydrogen: $H_2 + O^{2-} = H_2O + 2e^-$ On logistics fuel: $C_xH_{2y} + (2x + y)O^{2-} = yH_2O + xCO_2 + (x + y/2)e^-$	$\frac{1}{2} O_2 + 2e^- = O^{2-}$ $(x + y/2)O_2 + (x + 2y)e^- = (2x + y)O^{2-}$	Large (>1 kW) prototypes available. Tolerant to poisons. Most compatible with logistics fuels. Significant government and commercial investment.	High temperature management and corrosion. Start-up time for high temperature system. Fragility of ceramic system.

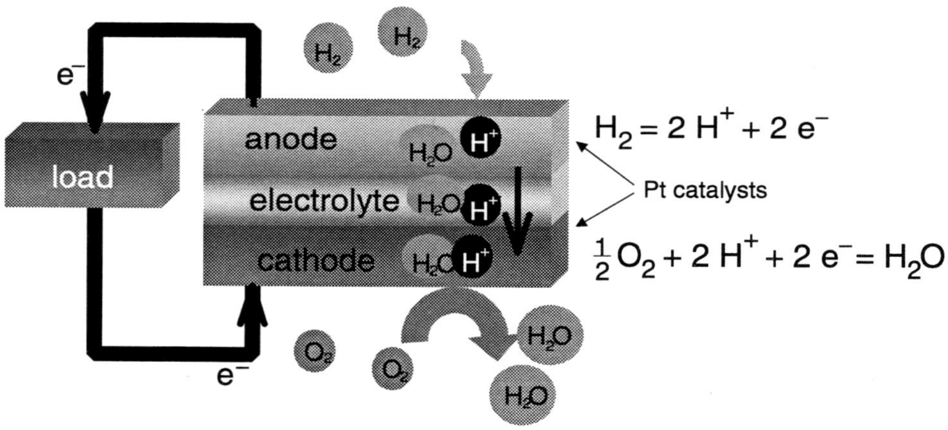

FIGURE D-2 Schematic of proton exchange membrane fuel cell. Hydrogen is catalytically oxidized by platinum at the anode to protons; the protons flow through the solid polymer membrane electrolyte to the cathode, where they reduce oxygen to water on platinum catalysts. The electrons from and to the oxidation and reduction reactions are forced through an external load. Additional components of the fuel cell are interconnects, current collectors, and sometimes gas diffusion layers. Multiple cells are combined to form a stack.

ideally be able to produce this valuable commodity for the soldier in the field. In spite of all of these advantages, however, the practical emergence of fuel cells has been delayed by material science challenges and by the lack of a mature technology and a supporting industrial base.

The selection of the fuel is critical to the building of high-energy portable power sources, given that the fuel is the source of the energy and the fuel cell is merely the conversion system. The specific energy and energy density of various fuels are listed in Table D-8. Hydrogen gas has the

TABLE D-8 Specific Energy and Energy Density of Various Fuels

Fuel	Btu/cf	Btu/lb	Wh/L	Wh/kg
Methane (g)	183,000	221,500	1,892	142,979
Methane (L)	183,000	570,000	1,892	367,937
Propane (L)[a]	724,000	19,937	7,485	12,869
Butane (L)[a]	695,000	19,678	7,186	12,702
Methanol	432,000	8,700	4,466	5,616
Ethanol	570,000	11,600	5,893	7,488
Ammonia	317,000	8,000	3,277	5,164
Hydrogen (g)[b]	56,000	52,000	579	33,566
Hydrogen (L)[c]	229,000	52,000	2,368	33,566
Gasoline	876,000	19,100	9,057	12,329
JP-8 (logistics fuel)	963,880	18,600	9,925	12,006

NOTE: Data are in terms of their lower heating value (LHV), or use at temperatures greater than the boiling point of water. For a DMFC operating at 60°C, the higher heating value (HHV) of methanol is 6,088 Wh/kg. The HHV of hydrogen gas is 39,504 Wh/kg at 15°C.
[a]Liquid at 27°C.
[b]Gas at 27°C and 3,000 psi.
[c]Liquid at cryogenic temperature (–253°C) and 1 atm.

highest specific energy of all fuels, but it also has the lowest energy density. The energy density of hydrogen gas improves if it is stored at high pressure (5,000 psi) or in metallic hydrogen storage alloys, as discussed below. Logistics fuels (such as JP-5 and JP-8) are the best choice for energy density, but the fuel must be reformed to hydrogen for portable proton exchange membrane (PEM) systems. There are pros and cons for each fuel cell that must be considered in the context of the particular fuel cell.

All the reactant, product, and thermal management functions of the fuel cell are accomplished with balance-of-plant (BOP) components/systems. Depending on the size and/or complexity of the fuel cell system, BOP components may be intimately integrated into the fuel cell stack or attached as distinct external components. BOP components can be energetically passive or require some parasitic power from the fuel cell stack to operate, so careful attention must go into fuel-cell design to achieve a high-efficiency system.

Proton Exchange Membrane Fuel Cells

Hydrogen PEMFCs are the simplest and most reliable type of fuel cell demonstrated to date. Hydrogen-fueled PEMFCs have been shown to be robust and reliable in real-world field tests, generating power at subzero temperatures up to normal operating temperatures of ~80°C. Work on hydrogen-fueled PEMFCs is currently receiving substantial government and commercial funding in the United States and abroad, as they have been identified as the best fuel cell for automobiles.

Stack development of PEMFCs is fairly mature as a result of large investments by the public and private sectors over the last 10 to 15 years. The electrolyte in PEMFCs is usually a perfluorosulfonic membrane (e.g., DuPont's Nafion), and new, lower-cost membranes are emerging (e.g., Polyfuel and Gore). The anode and cathode reactions are typically catalyzed by platinum at loadings of 0.2 mg/cm^2, but these loadings should decrease as research in this area progresses. The membranes and catalysts must be appropriately humidified, and their performance suffers when they become too dry or wet.

The most advanced portable PEM/H_2 systems use compressed hydrogen to simplify the fuel issue. But even when operated on pure hydrogen fuel, PEMFC systems require extensive control systems for optimum operation. Figure D-3 shows a mass flow diagram of a hydrogen-fueled, field-tested portable power system, the Ball Aerospace PPS-50 (TRL = 7). The fuel cell stack was obtained from H-Power, a now-defunct small company, although there are now other suppliers of 50-W fuel cells (e.g., Protonex and Neah). A complex BOP architecture is expressed for this system in order to provide greater versatility for end users wishing to operate the fuel cell system in as broad a range of environments as possible without retrofit. The PPS-50 fuel cell system is electronically controlled with a microcontroller utilizing various sensors for monitoring stack voltage, current, temperature, and so on. The system is designed to manage hydrogen delivery, oxidant air feed, cooling air for heat removal, and product water from the stack (Ball Aerospace, 2003).

The weight of the system is 2.9 kg, its volume is 4.26 L, and its demonstrated specific energy is 540 Wh/kg when running a 6 percent by weight hydrogen solution at 50 W for 72 hr (3,600 Wh at 6.6 kg). A 24-hour mission at 50 watts would have a specific energy of 286 Wh/kg (Ball Aerospace, 2004). Lower power (<20-W) hydrogen PEM

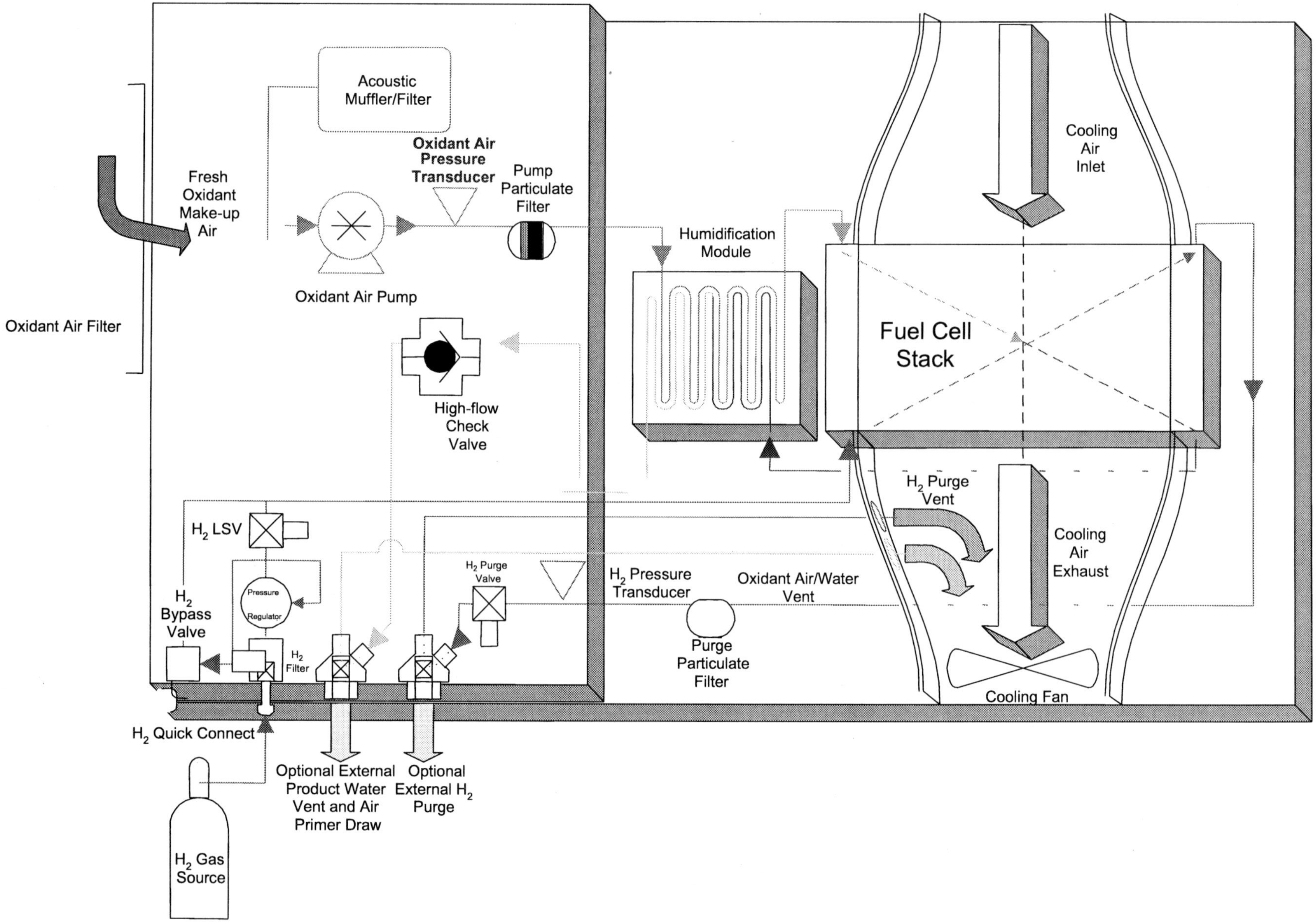

FIGURE D-3 Mass flow block diagram of a Ball Aerospace PPS-50 50-W hydrogen fuel cell system. SOURCE: Ball Aerospace.

fuel cell systems have also been developed by Ball Aerospace and others, thereby demonstrating the scalability and versatility of the technology.

Hydrogen Sources

The successful operation of a PEM/H_2 system is contingent on the identification and integration of a suitable hydrogen source. Whereas the PEMFC stack technology itself may be at TRL 7 to 8, the TRL levels of the hydrogen sources can be as low as 2, which drags down the readiness of the whole system. The integration issues, discussed below, keep some of the technologies from being implemented on a small scale for the soldier. The military must also consider any logistics or safety burdens that might be imposed by the use of the hydrogen source.

There are two types of hydrogen source: Type 1, hydrogen storage sources; and Type 2, hydrogen generation sources. Additionally, hydrogen sources are characterized by their need for fuel cell system resources such as electricity, heat, water, and so on. A Type 1 or Type 2 hydrogen source that is independent of the fuel cell system is termed a passive hydrogen source. Hydrogen sources requiring fuel cell resources (heat, electricity, water, etc.) are considered active and/or coupled. The characteristics of several active and passive hydrogen sources are detailed in Table D-9.

Hydrogen storage systems include pressurized tanks and metal hydrides. A pressure vessel is a passive hydrogen storage system where the hydrogen exists in its diatomic form (H_2) and can be extracted via a pressure differential and without the application of heat. Pressurized hydrogen is convenient and follows dynamic fuel cell loads. State-of-the-art composite pressure vessels with valve and regulator assemblies routinely demonstrate better than 5 weight percent (of system weight) H_2 gas storage when sized for large PEMFCs (e.g., ~50-kW systems for automobiles). Owing to the larger surface area:volume ratio in smaller systems and the relatively larger weight burden of the gas regulator, 5 weight percent hydrogen storage is not available for portable power sources, although it might be possible with development effort. Compressed hydrogen has been plagued by the perception that it is unsafe, but the Department of Energy is in the process of assessing safety issues, as recommended in a recent report (NRC, 2003).

Cryogenic hydrogen storage systems, such as the space shuttle's supercritical hydrogen storage tanks, demonstrate 29 weight percent H_2 and are active storage systems because heat must be applied to remove the hydrogen as boil-off gas.

TABLE D-9 Dependence of Select Hydrogen Sources on Fuel Cell Resources

	Specific Energy Ideal (Wh/kg)			Dependence on Fuel Cell Resources				
Energy Source	Net System[a]	Theoretical[b]	Fuel Processor Efficiency[c]	H_2O	H_2	Electric	Heat	Central
Liquid H_2 (PRSA at 29%)	4,340	9,477	0.96 active			x	x	x
HP H_2 gas (8,500 psi, 5.3%)	826	1,732	1.00 passive					
Metal hydride (2%)	312	654	1.00 active				x	
$LiAlH_4 + xNH_3 \rightarrow yH_2$ (AHHG)	2,071	4,343	1.00 passive					
$AlH_3 + \nabla \rightarrow Al + {}^3/_2H_2$ (thermal decomposition)	1,677	3,662	0.98 active			x		x
$C(SiH_3)_4 + 6H_2O \rightarrow 12H_2 + C(SiO_{3/2})_4$ (organosilane)								
With own H_2O	1,464	3,233	0.95 active			x		
With fuel cell H_2O	2,486	5,794	0.90 active	x		x		x
$LiH + H_2O \rightarrow LiOH + H_2$ (LiH hydrolysis with own H_2O)	1,209	2,537	1.00 passive					
With fuel cell H_2O	3,949	8,283	0.90 active	x		x		x
$LiAlH_4 + 2H_2O \rightarrow LiAlO_2 + 4H_2$ ($LiAlH_4$ hydrolysis)								
With own H_2O	1,665	3,564	0.98 active			x		x
With fuel cell H_2O	2,979	6,943	0.90 active	x		x		x
$2NH_3 + \nabla \rightarrow N_2 + 3H_2$ (NH_3 cracking)	2,213	5,802	0.90 active		x	x		x
$CH_3OH + H_2O + \nabla \rightarrow CO_2 + 3H_2$								
MeOH/H_2O reforming	1,600	3,948	0.85 active		x	x		x
With fuel cell H_2O	2,353	6,168	0.80 active	x	x	x		x
$CH_3OH + H_2O \xrightarrow{cat} CO_2 + 6H^+$ (MeOH in DMFC)	2,317	6,088	0.90 active	x		x		x

NOTE: PRSA, power reactant storage assemblies; AHHG, ammonia hydride hydrogen generator; HP, high pressure; DMFC, direct methanol fuel cell.
[a]Weight of containment, catalyst, coreagents, and converter/reactor plant not included in specific energy calculation. Net system energy yield assumes 100 percent chemical yield from reaction, with subsequent energy losses due to fuel cell voltage of 0.7 V, fuel cell system and generator balance-of-plant efficiency, and fuel utilization efficiency.
[b]Same assumptions as in footnote *a* except that all converter efficiencies are 1.0 and fuel cell voltage is 1.229 V.
[c]Impact on fuel cell balance-of-plant only.

Metal hydride storage systems are active because heat is required to desorb or release hydrogen from the hydride. Metal hydride systems operated at room temperature can release more than 2 percent by weight H_2, and those used at higher temperatures can release more than 4 percent by weight H_2. Rechargeable metal hydrides have limited cycle lifes of 100 to 1,000 charge/discharge cycles due to decrepitation of the metal, which causes packaging problems. Likewise, many high-performance composite pressure vessels and cryogenic storage systems also have limited cycle life and must be inspected and recertified regularly. As in the case of compressed hydrogen, the actual weight percent of hydrogen stored will decrease with the size of the hydrogen storage system.

Hydrogen generation sources include chemical hydride generators, fuel reformers, and electrolyzers, all of which produce hydrogen gas by a chemical reaction. Elements occupying the first three rows of the periodic table (before the transition elements) form hydrogen-rich storage compounds and are frequently considered practical fuels for hydrogen-generating systems. The choice of which hydrogen-bearing compound to employ when engineering a hydrogen source is based on the gravimetric and volumetric hydrogen density of the compound. Figure D-4 shows the gravimetric density of hydrogen in several light-element compounds, including compounds containing carbon, nitrogen, oxygen, silicon, and other light metal atoms, that are traditional candidates for hydrogen generators. The hydrogen-generating compound is selected not only on the basis of its energy content, but also on the basis of engineering considerations such as handling, storage, and mixing approaches and the releasability of hydrogen from these compounds with low energy penalty and low system complexity.

The reformation of oxygenated organic fuels such as methanol with water (a mildly endothermic process), the thermal decomposition of light metal hydrides such as AlH_3, and the cracking of ammonia all require some energy from either the fuel cell or the combustion of hydrogen. These sources of hydrogen are generally controllable at the expense of additional system complexity, cost, and—often—reliability. The reforming could be carried out in conjunction with a hydrogen separation membrane to yield pure hydrogen with no nitrogen dilution. The advantage of methanol and ammonia is that tank pressures are low. Ammonia has the additional attribute of having only hydrogen and nitrogen products, the latter of which would be rejected by a membrane.

Reforming hydrocarbons—a source of hydrogen for PEMFCs—is even more challenging. Because these fuel cells operate at low temperatures (60-80°C) and temperatures above 400°C are needed for hydrocarbon reforming,

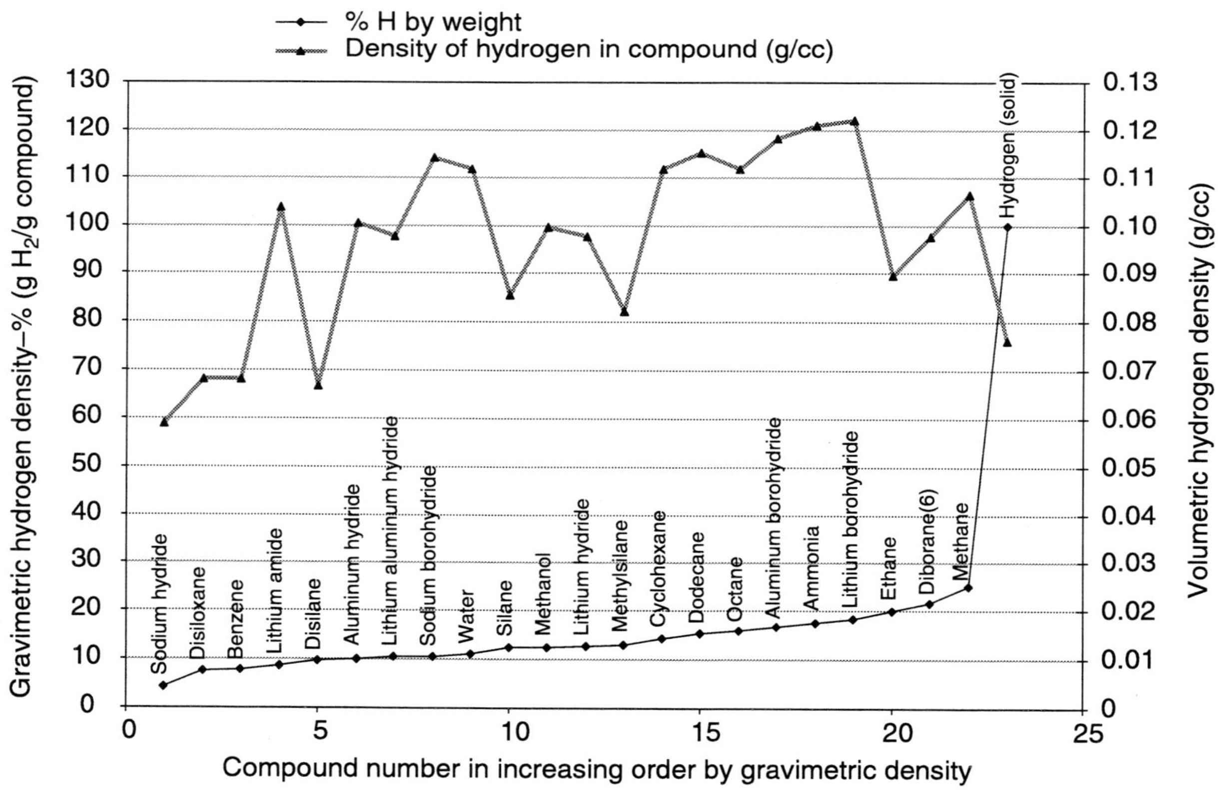

FIGURE D-4 Specific gravimetric hydrogen densities of select compounds.

an additional reformer is needed to convert hydrocarbon-based fuels to hydrogen. The additional reformer adds significantly to the weight and size of the small system. Water is required in reforming to shift carbon monoxide (CO) impurities to CO_2, both for human safety and to prevent the Pt catalysts from being poisoned or deactivated by the CO. Sulfur impurities, which are also deleterious to the Pt catalysts, must also be removed with sorbents. The requirements for fuel reforming were covered in detail by JASON (2003). Overall, reforming decreases the energy content of the fuel and adds considerable size and weight to the fuel cell system, so much so that the PEMFCs operating on logistics fuel were not considered to be a viable technology for portable power (see Chapter 2).

In the future, reforming for PEMFCs might be solved with compact microchannel or MEMS reformers, particularly for methanol solutions. Microchannel reactors are at the heart of a compact methanol steam reformer that has been developed by researchers at PNNL for the Army. The PNNL team has worked on microchannel reformers for a variety of fuels (Hu et al., 2003). A fully packaged microchannel reformer system for methanol, including CO cleanup, heat exchangers, and integration with a 150-W PEMFC stack, is now being evaluated by CECOM.

Recent research also reports that ethanol may be reformed at high conversion rates (Deluga et al., 2004), although this technology is presently too immature to make recommendations. Flammability and safety are the key issues that must be addressed for the DOD to consider alcohol-based fuels.

Direct Methanol Fuel Cells

DMFC technology avoids problems with fuel storage, because the catalysts at the fuel cell anode can directly oxidize liquid methanol, eliminating the need for complicated fuel storage components. DMFCs are a derivative of PEMFC technology in that they both use solid electrolytes (typically Nafion) and have similar Pt catalysts at the cathode side. At the anode, they utilize platinum-ruthenium (Pt-Ru) catalysts, which work for the methanol oxidation reaction in Table D-7. However, because the catalysts suffer from efficiency losses, or overpotentials, DMFCs typically operate at 0.3 V less than hydrogen-fueled PEMFCs (0.4 V vs. 0.7 V), making the thermodynamic efficiency of the DMFC cell chemistry low (32 to 40 percent). The practical efficiency is even lower as a result of the technological issues discussed below, among them methanol crossover.

Key technological issues in DMFCs are their high catalyst loadings, methanol crossover, and water management. DMFC anodes have 10-fold higher catalyst loadings per unit area than hydrogen-fueled PEMFCs. The cost of catalyst reflects the price of precious metal futures and makes devices delivering more than 100 W prohibitive in cost. Promising research is under way on how to replace the Pt-Ru catalysts with catalysts having little or no precious metal content, but a practical solution has not yet been developed.

Methanol crossover refers to the leakage of fuel from the anode compartment to the cathode. This is a serious problem, because the methanol that reacts with oxygen at the cathode produces no electricity, decreases the activity of the catalysts at the cathode, and creates an increased thermal burden on the overall DMFC system. Methanol and other low-weight alcohols cross through polymeric membranes because methanol resembles proton-water complexes (or hydronium ions) at the molecular level, allowing it to be dragged along with the ions conducting through the membrane. Strategies have been developed to use a selective gas diffusion layer (or other means) to control the rate at which methanol is supplied to the anode, thereby reducing or eliminating the amount of excess methanol that can permeate to the cathode (Grubb, 1970; Ren et al., 2001). Such approaches have led to fuel utilization rates of better than 93 percent, but they also limit the power density of the membrane or stack.

Alternatively, numerous researchers are investigating new membranes that are less permeable by methanol while retaining protonic conductivity. New polymers are being developed at Virginia Tech, PolyFuel, Inc., and the Gas Technology Institute,[9] several of which have been shown to conduct three times less methanol than conventional Nafion membranes (Hickner et al., 2002; Cooper and Cox, 2003). The new materials are still being evaluated for their ability to withstand long-term exposure to variations in methanol concentration, temperature, and current flux without degradation in performance, so their practical impact is not yet known. Membranes that are stable in high methanol concentrations and exhibit lower crossover will also allow the system to be stored at lower temperature without concern for freezing of the fuel cell stack (a 5 molar solution of MeOH freezes at about –10°C, while an 8 molar solution freezes at –21°C).

Water management is a concern because of the large amount of water involved in the DMFC reaction (see Table D-7). Extensive BOP is required to condense and store the product water and to mix pure methanol with the stored water to obtain methanol concentrations that are optimized for the membrane and stack (see Figure D-5). Product water management is further complicated by a phenomenon known as the electro-osmotic drag (EOD) of water across the membrane from the anode to cathode. Approximately 3 water molecules are dragged across the membrane for each proton conducted, so as many as 18 water molecules can be dragged from the anode to the cathode for every molecule of methanol oxidized. The rapid depletion of water at the anode could result in flooding of the cathode if active water recovery

[9]Found online at http://www.gastechnology.org. Last accessed on January 28, 2004.

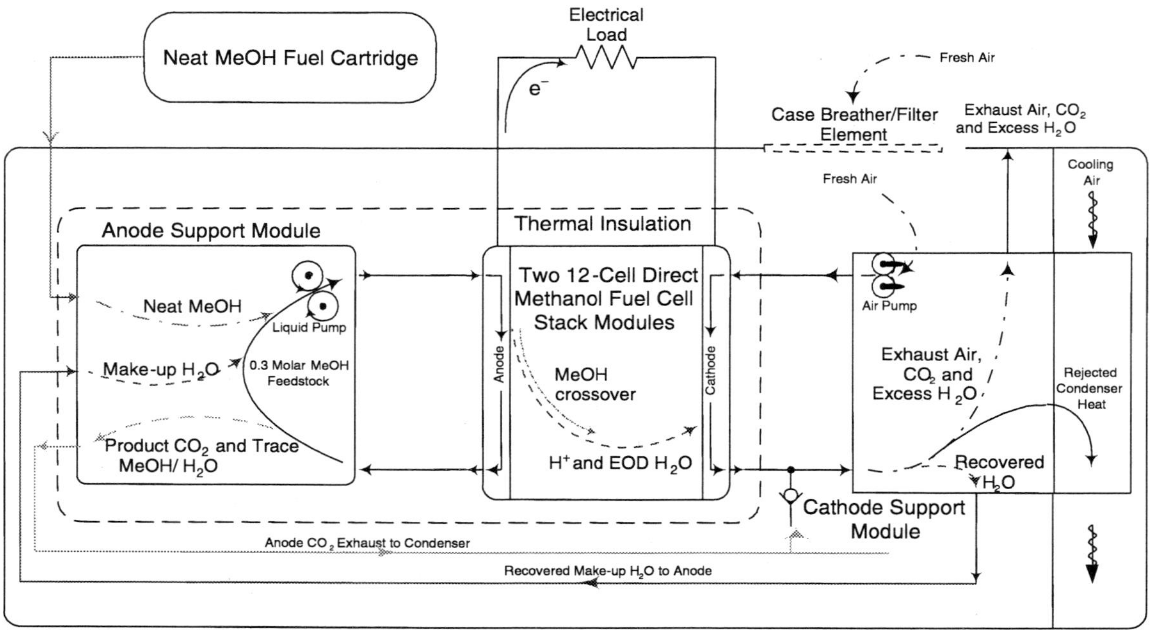

FIGURE D-5 Schematic of Ball Aerospace 20-W DMFC energy converter showing the active fuel and water management BOP subsystems used to maximize the specific energy yield of methanol above 2.0 Wh/g. SOURCE: Ball Aerospace.

from the cathode back to the anode is not practiced. Electro-osmotic drag has also been attributed to methanol crossover, so the lower the EOD, the smaller the methanol crossover. PolyFuel has demonstrated EOD coefficients of less than one (one water per proton) for MEAs based on its proprietary Z1 membrane and a corresponding reduction in methanol crossover to one-third that of Nafion (Cooper and Cox, 2003). Recently, novel low power (<5 W) DMFC concepts demonstrated by Mechanical Technology Inc. (MTI) MicroFuel Cells have broken this paradigm by demonstrating a passive laboratory DMFC operating on 100 percent methanol at the anode.

A 20-W DMFC system, the DMFC-20, has been developed by Ball Aerospace in collaboration with Los Alamos National Laboratory for the DARPA Palm Power program. The fully packaged system (TRL = 6) is now being tested by the U.S. Army and others. The diagram for the system is shown in Figure D-5. The system utilizes recirculated mixtures of 0.250 to 1.00 molar solutions of methanol (MeOH) and water. A fresh feedstock fuel mixture is circulated through the anode, sweeping out product CO_2 in the fuel cell anode exhaust. Careful systems design must consider mixed phases (gases and liquids) simultaneously managed by the system. Orientation-insensitive operation of the DMFC-20 is achieved, but it is complicated by the presence of liquids and gases in both the anode and cathode loops of the system. This DMFC-20 exhibits a specific energy yield approaching 2.0 Wh/g MeOH (0.33 (η_S) × 6.088 Wh/g), and improvements in the catalyst and membrane technology might increase the yield to perhaps 3.0 Wh/g MeOH.

The energy density of a fully integrated 20-W system is 540 Wh/kg when equipped with two fuel canisters, sufficient for a 72-hour mission. This operational configuration weighs 2.95 kg, has a volume of 2.93 L, and provides 1,600 Wh of energy. The energy converter alone weighs 1.75 kg and has a volume of 1.56 L. The DMFC-20 system with an 800-Wh fuel canister is shown in Figure D-6.

Additionally, Ball Aerospace has developed a 60-W DMFC, and Giner Electrochemical Systems has developed 50- and 150-W units. Smart Fuel Cells of Germany also reports the development of a portable DMFC.

Other DMFCs are being developed for operation at 0.5 to 2.5 W specifically as power modules for cell phones. In this power regime, both Polyfuel and MTI have demonstrated DMFC power source prototypes (TRL = 5-6) suitable for powering cell phones in the 0.5- to 2-W class. All of the 0.5- to 2-W systems in this power class are air-breathing and employ passive fuel management. Control electronics are typically employed to perform voltage boosting and internal battery recharging. The first Polyfuel prototype was fitted directly to the back of a Nokia cell phone and was able to power the phone in standby and receive mode.[10] An

[10]Polyfuel demonstrated a working prototype to committee member Jeff Schmidt at Ball Aerospace and Technologies in 2002.

FIGURE D-6 A DARPA/Ball Aerospace and Technologies operational DMFC-20 (20-W direct methanol fuel cell with a hard-packaged 500-cc fuel canister). SOURCE: Ball Aerospace.

onboard secondary battery or ultracapacitor is used to provide peak power (2 to 3 W for transmit mode). Power systems of this type are not presently part of PolyFuel's business model but serve to demonstrate the use of its alternative non-Nafion membrane for DMFC applications. MTI MicroFuel Cell,[11] however, has adopted a business model for the commercialization of small 0.5- to 5-W DMFC power sources for cell phones and small personal electronics such as personal digital assistants (PDAs). Several of the company's prototypes are displayed on its Web site at http://www.mtimicrofuelcells.com/technology/prototypes.cfm.[12] MTI demonstrated several of these systems at the 2003 fuel cell seminar, held November 3-7, 2003, in Miami Beach, Florida. The demonstrated 0.5-W continuous and 2-W peak operational MTI prototype is 0.09 L and weighs ~0.09 kg (Acker, 2003). The purported energy yield from pure methanol is 1.0 Wh/cc MeOH (1.25 Wh/g MeOH); however, the current systems operate on a 50 percent water/methanol mixture stored in small 20-cc fuel cartridges. For a 72-hour mission, four of these cartridges would be required for 40 Wh of energy (36 Wh for the mission). The total 72-hour mission system weight is calculated to be 0.165 kg, yielding a specific system energy of 242 Wh/kg (also 242 Wh/L, as this system is about as dense as water). MTI has also demonstrated a larger 5-W DMFC configured in the form factor of a BA5590 size battery of 0.9 L (2 in. × 4.4 in. × 5 in.) (Acker, 2003). The system also includes a secondary battery able to provide 50-W peak power surges for an undisclosed duration. This system employs an active air mover (fan) for cathode air handling and cooling. The converter weight is estimated at 1 kg and includes an internal 100-cc, 50 percent water/methanol mix fuel cartridge. The energy yield from this unit is also purported to be 1.0 Wh/cc MeOH. The specific energy estimate for a 360-Wh mission (72-hour mission at 5 W average power) is 212 Wh/kg, requiring eight fuel cartridges (400Wh/1.88 kg). MTI announced in an October 2003 press release that it had achieved passive operation on 100 percent methanol at the anode (MTI, 2003). This allows neat methanol to be used in the fuel cartridges, thereby increasing the specific energies given above by as much as 50 percent.

The committee also obtained data from Motorola Labs on performance specifications for its 2.5-W DMFC.[13] The

[11]Found online at http://www.mtimicrofuelcells.com. Last accessed on January 28, 2004.

[12]Last accessed on January 28, 2004.

[13]Data provided to committee members Jeff Schmidt and Karen Swider Lyons by Jeanne Pavio, Manager, Fuel Cell Development, Motorola Labs.

energy converter operates on pure methanol, yielding ~0.8 to 0.9 Wh/g MeOH, and weighs 350 g without fuel. Here, too, the energy converter density is about 1 g/cc, so the converter volume without fuel is ~350 cc. The calculated specific energy of this system for a 144-Wh mission (72 hr at 2 W) is 262 Wh/kg (144 Wh/0.55 kg).

The efficiency of the low-power (5 W or less) DMFCs is less than that of higher power (20 W and higher) systems, because the former typically use passive BOP components and the latter use active BOP management. The smaller DMFC systems usually have passive air cathodes and noncirculating liquid, or a small fan that passes air over the cathode to help evaporate product water and replenish oxygen. The fuel is a premixture of methanol and water supplied by a small cartridge mechanically pressurized by a spring or elastomer. These low-power systems operate at reduced temperatures, thereby lowering the operating voltage of the cell, usually by 0.3 V or less. The use of premixed fuels, the lower voltages of cell operation, and the need for voltage-boost electronics reduce the overall efficiency of the system. Hence the specific energy yield of the <5-W DMFCs is typically less than 1.2 Wh/g when pure methanol is the fuel and less than 0.6 Wh/g when premixed methanol is the fuel. Advantages of the low-power, passive DMFCs are their limited complexity (eventually translating to lower cost) and quietness.

By comparison, the energy yield from a fully active DMFC-20 is typically between 1.5 and 2.0 Wh/g on pure methanol. The two- to threefold improvement in specific energy yield in the Ball Aerospace DMFC-20 system is largely due to the system's active management of reactants and products by BOP components. At larger sizes, BOP reactant and product management become more efficient overall, and they become smaller relative to the overall system size and weight. BOP components for >20-W systems are commercially available or could be developed within a relatively short time, making them immediately suitable for systems development and integration. In the future, it is possible that BOP components that can improve the specific energy yields of smaller systems may be realized by MEMS technologies or mesoscopic machines/devices.

Solid Oxide Fuel Cells

Solid oxide fuel cells (SOFCs) have been in development in the United States, with support from DOE, since the 1960s, predominantly for use as terrestrial power plants. SOFCs were not considered as a possible power source in *Energy-Efficient Technologies* (NRC, 1997). However, in the last few years, several R&D efforts, many of them under the DARPA Palm Power program, have focused on developing man-portable SOFCs for military and commercial applications (1 to 100 W). The successful operation of SOFCs is dependent on robust materials and cell designs. A general review of the findings can be found in: *Ceramic Fuel Cells* (Minh, 1993). DOE's *Fuel Cell Handbook* also gives a thorough review of SOFCs (DOE, 2002).

The clear advantage of SOFCs as portable power sources for the military is their ability to operate on hydrocarbons with little or no reforming. Because they operate at high temperatures (600-800°C), one can take advantage of internal reforming, whereby the fuel is oxidized by the reactants at the fuel cell anode. This section focuses on issues faced by the operation of small SOFCs on logistics fuels (JP-8) or other hydrocarbons.

Hydrocarbon fuels can be oxidized to CO_2 and H_2O at the SOFC anode, as summarized in Table D-7. If there is insufficient oxygen for complete oxidation of the hydrocarbon, CO may form, which may also be used as a fuel:

$$CO + {}^1/_2O^{2-} = CO_2 + e^-$$

At high temperatures, particularly in the presence of metal catalysts, CO undergoes reversible disproportionation to coke (solid carbon) and carbon dioxide:

$$2CO = C(s) + CO_2$$

Coke formation is possible either at the SOFC anode as hydrogen and fuel are consumed and the gas equilibrium changes, or in the fuel exit lines as an unreacted fuel in the exhaust cools. If coke is not controlled, it can adversely affect the performance of the cell by blocking catalyst surfaces and fuel passages. Coke formation is often prevented by adding water to shift the reaction to CO and H_2:

$$C(s) + H_2O = CO + H_2$$

For military systems, it would be ideal to not have to carry additional water for operation of the fuel cell, which would significantly penalize the energy density of the system, so methods for the suppression of coke formation are critical. It may be possible to efficiently recover some water from the anode exhaust (see Table D-7), but steam:carbon ratios of 2 or 3 are typically required to prevent coking, necessitating an additional water source when heavy hydrocarbons are reformed. Industrial processes often use steam:carbon ratios of 5 to 8 to extend the lifetime of the reforming catalysts. A possible solution is to carry a small amount of excess water, which can be used for internal reforming and then recuperated. This approach is being exploited by Altex Technologies Corporation to produce an efficient reformer for JP-8 (Ball Aerospace, 2004).

Long-chain hydrocarbons, in either the feed or the exit gas, are also prone to decomposition into coke owing to thermal degradation, and it is not clear if JP-8 can be successfully fed directly into the hot fuel cell. JP-8 can be converted to lighter hydrocarbons, which are less prone to thermal decomposition into coke, either by steam reforming or in a catalytic partial oxidation (CPOX) reforming unit. CPOX reforming units typically have Rh-based catalysts and operate at ~700°C. The CPOX units are compact and lightweight, but the overall system suffers from a drop in system efficiency because part of the fuel is oxidized in the CPOX unit before it enters the SOFC. As a result, it is diluted with nitrogen from the air and has less energy content.

Fuel processing might also be accomplished for future portable SOFCs with MEMS-based microchemical systems. Army-funded MURI programs on high-temperature microchemical systems for fuel reforming are being carried out at the Massachusetts Institute of Technology and the University of Illinois at Urbana-Champaign (UIUC). At MIT, novel microreactors and heat exchangers are being created for fuel combustion and reforming. Researchers recently reported on a microchemical reactor for butane and ammonia processing (Arana et al., 2003). The UIUC program has developed robust alumina microburners (Raimondeau et al., 2003) and penny-sized reactors that can produce the equivalent of 40 W of H_2 from NH_3 (Paur, 2003). If such systems can be proven reliable, efficient, and inexpensive, they may be useful as lightweight reforming components. Work by the University of Pennsylvania (Park et al., 2000) has shown direct oxidation, or internal reforming, of various long-chain hydrocarbons on the laboratory scale using new ceria/copper catalysts. The hydrocarbons are fed directly into the SOFC anode, where they are oxidized, eliminating the need for a CPOX unit to break down the hydrocarbons and water for internal reforming. The development of these systems is still preliminary, and the catalysts may face stability problems at temperatures over 700°C. A CPOX unit may still be needed with JP-8 fuel, as it may not be possible to feed the heavy hydrocarbons into the SOFC without thermal decomposition. More research is needed to determine whether the laboratory observations can be scaled into a practical military system. A system operating via direct oxidation on logistics fuels would have a clear military advantage as it would have no need for a prereformer or reformer.

Metal dusting of stainless steels, or corrosion of Fe- and Ni-containing materials due to the formation of metal carbides, may also be a failure mechanism in compact SOFCs operating on logistics fuels, as it can occur in gas streams rich in carbon monoxide and hydrogen between 425°C and 815°C.

Lastly, the sulfur in logistics fuels may react deleteriously with the catalysts in the SOFC anode and decrease performance. Although SOFC anode catalysts can tolerate up to 50 ppm of sulfur in the gas stream, logistics fuels typically contain more than 10,000 ppm sulfur. There are several options for solving the sulfur problem: The sulfur tolerance of the anodes can be improved, the sulfur can be removed by adsorption or scrubbing methods, or the SOFCs may be designed to operate on prepackaged, sulfur-free fuels. Sulfur-tolerant anodes would be ideal but have remained elusive to date. Adding a sulfur sorbent to the system is practical but increases the weight and complexity. The removal of sulfur with sorbents is being explored by researchers at Penn State, in collaboration with Altex Technologies Corporation. Use of a prepackaged fuel may be ideal for SOFCs but introduces logistics issues similar to those for primary batteries, albeit less severe, since there is a 10-fold higher energy density in the fuel compared with batteries.

Small SOFC systems do not enjoy the high efficiencies (>50 percent) reported for large SOFC systems, because the heat produced from the hydrogen oxidation and oxygen reduction reactions is insufficient to maintain the heating temperature of the fuel cells, necessitating the burning of fuel to keep the fuel cell stack hot. For a 20-W system, for example, 50 W of thermal energy might be needed to maintain the SOFC stack at 800°C. However, the heat generated by the electrochemical reactions might be only 25 W, leaving a shortfall of 25 W. This might be met by burning unutilized fuel, but it might also require the burning of additional fuel. Therefore, it is best to assume practical efficiencies on the order of 30 percent for small SOFCs operating on a liquid hydrocarbon fuel. If a CPOX unit can be avoided, for instance with direct oxidation, the system efficiency might be higher (~35 percent).

SOFCs targeted for operation at 2, 20, and 100 W and 1 to 3 kW are likely to have vastly different designs. Devices to produce 2 W cannot be fabricated except—possibly—using a MEMS-type design. Such an approach is the subject of an Army-funded multiuniversity research initiative (MURI) program at MIT (TRL 1 to 2) and is being pursued by Lilliputian Systems of Woburn, Massachusetts. The development of MEMS-based SOFCs is very high risk, because issues such as high-temperature seals (for managing the thermal mismatch between the silicon-based MEMS structure and SOFC materials) and the thermal engineering are critical. It is also not clear if long-term SOFC operation can be achieved, given the possibility of failure of seals and/or poisoning of the YSZ electrolyte by silicon from the MEMS fuel-cell frame and traces of silicon from the fuel reformer.

Several 20-W SOFC designs are under development for the DARPA Palm Power program, and it is apparent that the designs and beliefs that are upheld for large terrestrial systems apply differently to these mesoscale systems. Note that portable SOFCs are still in the early demonstration stage (TRL 2 to 4), so much of the discussion below is based on projections.

The 20-W SOFC discussed in Chapter 2 is the microtubular-based system by Adaptive Materials, Inc. (AMI) of Ann Arbor, Michigan. Attributes of the system are summarized in Table D-10. The fuel cell stack comprises microtubules about 1.7 mm in diameter and 12-14 cm in length. The active SOFC part of each tube is 6 cm long, with several millimeters of the tube being dedicated to an integrated CPOX unit for breaking down the fuel (butane) to light hydrocarbons. The gas inlets for the tubes are at 200°C, enabling low temperature seals. The microtubular design is

TABLE D-10 Characteristics of Butane-Fueled 20-W Solid Oxide Fuel Cell System by Adaptive Materials, Inc.: Breadboard Versus Projected Attributes

Characteristic	Laboratory Achieved, Breadboard	3-Day Mission, Projected
Specific energy	340 Wh/kg (10 days)	1,000 Wh/kg
Power density	1.8 W/cm^2 of tube	
Specific power	4.7 W/kg (wet); 200 W/kg (stack)	13.9 W/kg (wet)
Fuel	Butane, 2.0 kg	Butane, 0.4 kg
Signature		Thermal exhaust plume less than soldier's breath
Cost (initial) per Wh	Not known	Not known
Fuel consumption	27.8 g/hr	6.25 g/hr
System weight, dry	2.3 kg	1.0 kg
System weight, wet	4.3 kg	1.45 kg
Efficiency at rated power	6% (20-W continuous load)	29%
Altitude impact	85% at 20,000 ft (demonstrated)	>90% at 15,000 ft
Form factor		4 × 4 × 8 in. (approximately)
Life/cycles	Demonstrated multiple start/stop cycles	Minimum 250 start/stop cycles
Maintenance		Minimal, simple module replacement
Environment	No underwater	No underwater
Orientation dependence		None
Shock/vibration		100% survival after 10 ft drop onto concrete
Technology readiness level	4	
Start-up time	<3 min	
Fuel utilization	85%	

also resistant to thermal shock, allowing the AMI system to be started in 1 to 3 min. As of February 2004, AMI had demonstrated a packaged 20-W SOFC prototype system that is thermally self-sustaining on butane fuel. The demo systems comprise a fuel cell stack, insulation, recuperator, electronic controls, and a battery (TRL = 4). The exhaust temperature of this packaged system is 40 to 50°C, and the package has survived 40 G drop tests (Crumm, 2004). It can also be stopped and started multiple times with no detrimental effects on performance. The system still suffers from inadequate thermal insulation and heat recuperation, so the power efficiency is presently about 13 percent and the energy density is 510 Wh/kg. With improvements in the BOP and systems engineering, AMI expects to build stacks with a rated specific energy of 1,000 Wh/kg for a 72-hr mission and 2,500 Wh/kg for a 240-hr mission.

SOFC systems of 75 to 200 W would be attractive as battery chargers if they could be run directly on logistics fuels. For these intermediate-sized systems, planar designs might be more practical than microtubular designs, as the active area of the SOFC plates can be somewhat enlarged while minimizing the resistance from interconnects. MSRI of Salt Lake City, Utah, has successfully demonstrated a breadboard ~100-W planar SOFC that operates on hydrogen, methanol, and ammonia. As developed, it was too heavy to be used as a portable power source, but it is now being integrated into a lightweight 75-W system by Mesoscopic Devices of Broomfield, Colorado. Operation on a sulfur-containing logistics fuel has not yet been demonstrated. Like the AMI system above, the MSRI planar stack can also be stopped and started with little decrease in the fuel cell performance; unlike larger SOFCs, small fuel cells can withstand thermal stresses.

SOFCs may also be developed as 1- to 3-kW systems. ITN Energy Systems estimates that a 5-kW system would weigh 15.5 kg dry and 118 kg with enough fuel for 72 hr (based on 40 percent net system efficiency). The dry weight of the system is competitive with PEMFC technology, as shown in Chapter 2, and steps could be taken to further reduce the weight of the stack. A hydrocarbon-fueled 2-kW SOFC is commercially available from Acumentrics,[14] but it has not been optimized for compactness.

SMALL ENGINES

Energy-Efficient Technologies for the Dismounted Soldier (NRC, 1997) considered small engines as a potential energy source for the dismounted soldier but did not consider them for specific applications. In general, the conclusions

[14]Found online at http://www.acumentrics.com. Last accessed on January 28, 2004.

drawn in that report are still valid, and the following is intended to upgrade the material in that study.

Small Internal Combustion Engines

Numerous small engines in the commercial sector may be adaptable to military needs. These range from hobby engines such as those used by model airplane enthusiasts to the more common types used in gardening tools such as leaf blowers and string trimmers. There are also a variety of advanced engine concepts, such as the micro internal combustion swing engine being developed at the University of Michigan (Crumm, 2004).

A survey of manufacturers' technical data for hobby engines (see, for example, OSengines.com and hobbyhobby.com) shows that the best engines have a specific power of 2 to 3 kW/kg for both two- and four-cycle engines, with the four-cycle engines being slightly heavier. In general, fuel consumption is 0.3 to 0.4 kg/kWh. Some manufacturers are introducing electronic fuel injection to produce an engine with lower fuel consumption. There is no data in the technical literature on the reliability of such engines; in general, hobbyists use hobby nitro fuel with 10-15 weight percent oil.

In the small engines developed for the consumer market, both two- and four-cycle engines are available. In general, two-cycle engines are about 10-15 percent efficient. Four-cycle engines are up to 25 percent efficient for mechanical energy. Both routinely use regular unleaded gasoline and some have been converted to run on other fuels. The specific power of these engines is about 0.45 kW/kg, with specific fuel consumption 0.2-0.3 kg/kWh, depending on operating parameters, among other things. Considerable data on reliability exist for these engines, and thousands of hours of life are possible with routine maintenance.

A variety of fuels can be used, ranging from methanol to diesel. Several companies, such as D-Star, Eagle Development,[15] and Foster-Miller, Inc.,[16] are taking hobby motors and motors designed for consumer tools, both two- and four-cycle versions, and converting them to run on the diesel cycle. In general, the fuel is a mixture of ether and kerosene. However, D-Star is using an atomizer to precondition the fuel and has successfully run converted engines on JP-8 for several hours. The Naval Surface Warfare Center Carderrock Division has run a Davis engine for longer than 20 hours on JP-8, with little or no engine fouling. There are insufficient data from these tests to estimate reliability.

To be usable for soldier power, suitable electrical generators must be used. In general, small permanent-magnet alternators have specific power of about 140 W/kg. Lightweight aircraft alternators have specific power of about 200 W/kg for units in the 500-W class. Both types can be over 80 percent efficient in converting kinetic energy to electricity. Taking the converter and engine efficiencies together, it is possible to achieve a maximum system efficiency of 20 percent or more from fuel to electricity.

Small engines typically have problems with durability and reliability. Use of battlefield logistics fuels presents problems for all small engines. In many cases heavily loaded bearings and sliding surfaces require lubrication. Logistics fuels typically have low lubricity, and this requires the use of lubricating additives. These additives seldom burn completely, so they contaminate the engine exhaust with noxious products. Deposits are likely to form on internal engine parts and on surfaces near the engine exhaust.

Creating good high-pressure gas sealing is also a problem for these small engines. Cylindrical surfaces appear to work best for sealing small engines.

Engines that have special lubrication systems have demonstrated the best durability and reliability. Minimal progress has been made in addressing these problems over the years, and there appears to be little hope of ever solving them.

MEMS-based Combustion Engines

Microturbines, or MEMS-based, micro gas turbine engines, were identified in *Energy-Efficient Technologies* (NRC, 1997) as a promising source of portable power for the Army of the future. The idea for such microturbines was conceived at the Massachusetts Institute of Technology, and that institution remains the leader in research in this area (Epstein et al., 1999). The development of MEMS microturbine systems was funded by the Army under a MURI program and is now funded at MIT under the Army Collaborative Technology Alliance.[17] Similar efforts to realize centimeter-sized gas turbine generators in the 10- to 200-W range have been started by Honda (Japan), IHI (Japan), the University of Tokyo (Japan), ONERA (France), the Singapore Institute of Manufacturing Technology (Singapore), and the Katholieke Universiteit at Leuven (Belgium). These groups are using a variety of fabrication approaches ranging from silicon/silicon carbide micromachining (MIT and Singapore), to ceramic injection molding/sintering (Honda and IHI), to conventional metal machining (University of Tokyo, KUL). The growth of research in this area is evidenced by two recent international symposia: Power-MEMS 2003 (December 4-5, 2003, Chiba, Japan, www.getinet.org) and ISMME2003, the International Symposium on Micro-Mechanical

[15]Found online at www.davisdieseldevelopment.com. Last accessed on January 28, 2004.

[16]Found online at www.Foster-miller.com. Last accessed on January 28, 2004.

[17]Found online at https://www.ctapower.org. Last accessed on January 28, 2004.

Engineering (December 1-3, 2003, Tsukuba, Japan, http://shourai.hitachi.co.jp/ismme2003/ISMME.html#English).

Numerous technical issues must be addressed before an actual MEMS microturbine system can be realized. There are materials issues, because the MEMS microturbine operates at 800°C, which is well in excess of a practical temperature for the long-term use of Si. Emerging MEMS fabrication methods must therefore be applied to achieve more rugged materials, such as SiC or SiN. In this new size realm, the dimensions of every aspect of the microengines must be modeled and tested. A new magnetic generator must be developed to convert the engine's motion to electricity. There are also critical issues with tribology and identifying bearings and a bearing design that can withstand the 1 to 1.4 million rpm of the microengine blades in the engine and the generator (Ehrich and Jacobsen, 2003). Near-term plans are for testing the engines with hydrogen, and much work is needed to run them on a hydrocarbon-based fuel.

Energy-Efficient Technologies (NRC, 1997) estimated the specific energy of a MEMS microturbine system to be 4,000 Wh/kg, but the MIT team has since changed its focus and is working on a 5 percent efficient system with a goal of 700 Wh/kg (Epstein et al., 2003). According to the formulas at the end of Appendix C and Figure C-4, any 5 percent efficient system will have a specific energy of less than 600 Wh/kg for a 1,440-W mission (72 hr at 20 W). Operating for 72 hr, a system weighing 250 g (excluding fuel) that operates at 5 percent efficiency would have a specific energy of 540 Wh/kg, and a 500-g system would be at 496 Wh/kg when operated for 3 days. To meet aggressive long-term goals of an order of magnitude increase in the energy of a battery, or approximately 2,000 Wh/kg, the system would have to run at >20 percent efficiency (see Figure C-4).

In addition to the challenges associated with system efficiency, the success of the MIT and other microengine programs hinges on the critical demonstration of self-sustainability for a gas turbine operation. Such a demonstration would indicate that a power plant was viable provided that problems with materials, bearings, and component efficiencies are solved. Until the projects are able to demonstrate the capability of a free-running micro gas turbine engine or a net positive engine output, they will remain at the TRL 1-2 stage in their development.

Refinement of components needed for efficient operation of the microengines also face critical challenges. Several components needed in a gas powered microturbine have been successfully demonstrated including a micro-scale high-speed compressor impeller and high-speed gas bearings (Johnston, et.al., 2003; Ehrich and Jacobson, 2003).

In the absence of a MEMS device that can operate in a self-sustained mode, fluid mechanics simulations would be required to evaluate the real performance potential and to estimate from the aggregation of components in the MIT system whether sufficient progress has been made toward achieving required component efficiencies. Such simulations are routinely made on larger sized turbo machinery to improve the component efficiencies by running system components on test rigs with very capable diagnostic flow measurement tools and thereby ascertain the real flow conditions within the machine. The diagnostic process is important because three-dimensional fluid flow analysis often fails to predict subtle but important effects within the machines, such as flow separations from the constraining walls. Such deviations from desired flows often greatly degrade component performance.

Detailed simulations, combined with three-dimensional fluid flow analysis, have allowed the optimization of larger machines to near-theoretical efficiency levels; however, the application of such simulations to microturbine components appears to be a very difficult challenge. The small size of the components makes application of existing diagnostic tools difficult if not impossible. If tools for this combination of analysis and design modification are not created, the process of component improvement may not proceed at a reasonable pace. The development of such tools, to be used in concert with existing analytical capabilities, will be time consuming and may be very expensive. Thus, the continuing improvements in component efficiencies that are required to make the microturbine generator a viable power source are likely to be very expensive and to have a limited chance of success. Because of the invention still required to achieve a coherent system design, the committee feels that the TRL of the MEMS microturbine technology is no higher than 2, and it did not consider the technology in Chapter 2.

Stirling Engines

External combustion engines such as steam engines and Stirling cycle engines were used in practical applications as long ago as 1800 but have since been largely relegated to history, except in a few embodiments, because of more efficient alternatives—namely, efficient internal combustion engines and electrical power from an ever-expanding grid. The primary advantage of the Stirling cycle is that the thermal process is steady state, which allows combustion optimization and energy recuperation. Further, steady-state combustion inherently has a lower acoustic signature than internal, impulsive combustion engines. For free-piston versions of Stirling engines, it is possible to operate two separate engines such that all vibration is canceled, resulting in an extremely quiet system. Early versions of Stirling engines employed exotic materials and had low specific power even though they were efficient converters of thermal energy to electricity. In recent years, however, advances in materials have resulted in components with high-temperature properties favorable enough to provoke interest in Stirling technology as a viable energy converter for some applications. It is currently a viable candidate for deep space exploration (http://www.grc.nasa.gov) and shows promise for battlefield and commercial applications (U.S. Army, 1993).

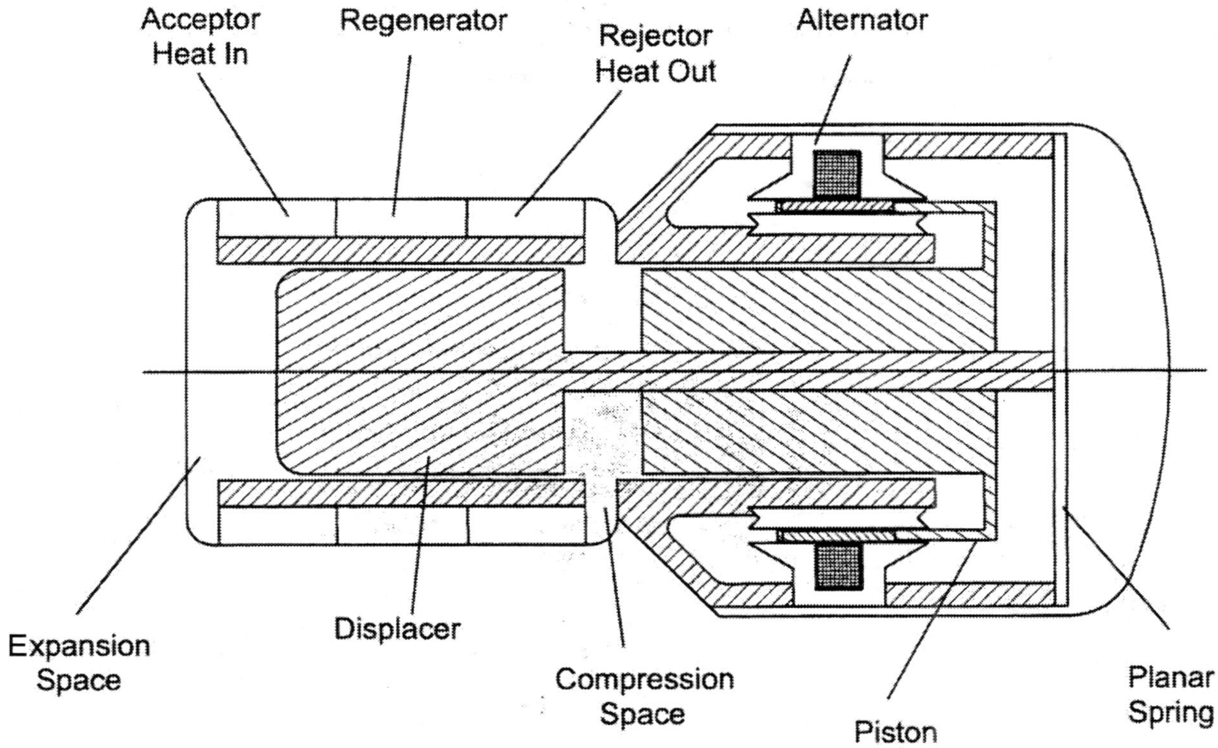

FIGURE D-7 Free piston Stirling engine showing component parts. SOURCE: SunPower Corporation.

Even though the technology has important desirable attributes, *Energy-Efficient Technologies* (NRC,1997) did not consider external combustion engines as viable candidates for soldier power owing to their low specific power, and to date, their primary commercial and military application is for cryogenic coolers.

The engine works because it is possible to alternately heat and cool an enclosed working fluid from a continuous flow external burner. The engine has five primary components: two pistons (or a piston and displacer), a regenerator, and two voids, or closed volumes, into which the gas and pistons expand (see Figure D-7). The regenerator section is a heat exchanger that alternately absorbs and releases heat. One of the volumes is maintained at a low temperature and is the compression space. The pistons are used to change the cylinder volume and to shuttle the working fluid back and forth. Work in the Stirling engine is generated by compressing and expanding the working fluid at different temperatures. The choice of working fluid is critical for the efficiency of the Stirling engine. While hydrogen would be the best working gas, it is difficult to contain in a sealed volume, so the most widely used gas is helium. There are several geometrical arrangements for Stirling engines. The free-piston version uses fluid forces to move the components, which results in no mechanical linkages to the piston or displacer. When coupled to a linear alternator, the device may use flex bearings or gas bearings, both of which allow the engine to be hermetically sealed into one compact unit with the fewest possible moving parts, none of which are in physical contact.

The most advanced Stirling engine that is likely to be applied for soldier power is the engine being developed by Sunpower, Inc. (Wood and Lane, 2003). The unit is currently being developed in a NASA Phase I SBIR for deep space applications. A recent visit to Sunpower confirmed the status of the system, which appears to be on the road to meeting or exceeding all specifications. The initial tests of the unit were at 31 W and a conversion efficiency of 29 pecent from thermal input to the head to electricity out.

As a general rule of thumb, one must at least double the mass of engine and converter for the remainder of the ancillaries needed to produce a working engine. Based on that figure, one could expect to produce a 35-W system with a dry weight of less than 1 kg. Figure D-8 is a conceptual drawing of a 20-W engine powered by a liquid propane energy source. It shows the relationship of the burner, heater head, and engine/alternator. It has a finned cooling system and would probably need forced air for cooling. This was

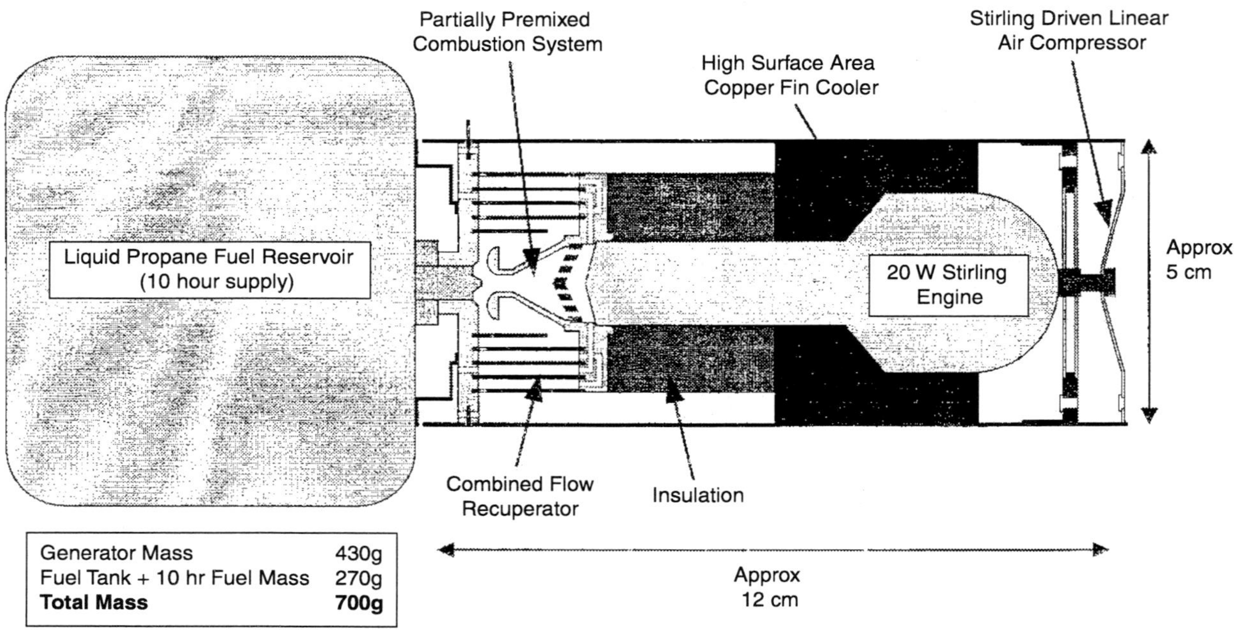

FIGURE D-8 Conceptual layout for a 20-W Stirling power system for soldier applications. Note that the unit is liquid-propane powered. SOURCE: SunPower Corporation.

used as the basis for the system projections reflected in Chapter 2 (see Tables 2-3 and 2-4). The unit has been resized for 1,440 Wh and the use of JP fuels.

Figure D-9 shows a 1-kW Stirling engine recently purchased by Auburn University as part of its hybrid electric program for silent watch applications. It was made for the cogeneration market in Europe and has been tested for thousands of start-stop cycles.

It is unclear whether the advances made in specific power for small engines will also translate to larger sizes; however, if they do, the pressure vessel and linear alternator mass could be reduced significantly to raise the specific power from approximately 44 W/kg to approximately 100 W/kg. Reliability is one of the most desirable characteristics of Stirling engine technology. In various embodiments, key components and full systems have been tested for thousands of hours (Schreiber, 2001). The longest trial cited by Schreiber—a testbed for the key flexbearing in the unit—ran for 12.6 years.

ADVANCES IN OTHER AREAS

Advances have been made in the areas of thermoelectric energy, thermophotovoltaics, and energy harvesting that make these areas candidates for soldier energy sources.

Thermoelectric Energy

Energy-Efficient Technologies (NRC,1997) considered thermoelectric (TE) power systems not viable for soldier power applications owing to their inherently low conversion efficiency and low specific power. TE devices utilize specialty materials that are able to convert a thermal gradient to electricity. The systems typically require a large heat gradient for maximum efficiency. The maximum specific power for converters mentioned in the 1997 study was from 15-20 W/kg, with conversion efficiencies less than 10 percent. Such efficiencies are typical of what is available commercially through firms such as Global Thermoelectrics,[18] whose generators range from 15 to 500 W and in general have system specific power in the range 0.7 to 0.9 W/kg. The fuel is natural gas or propane and consumption is about 0.2 kWh/kg. Total system efficiency is about 2 percent. If a recuperated burner is used, the total efficiency could be improved to approximately 8 percent. The low efficiency (<10 percent) of these systems makes them inherently difficult to integrate into high-energy-density power sources, because they must be very lightweight, as noted in Figure C-4.

The efficiency of TE systems is directly related to the properties of the TE materials, so there has been consider-

[18]Found online at www.globalte.com. Last accessed on January 28, 2004.

FIGURE D-9 1-kW Stirling engine recently purchased by Auburn University as part of its hybrid electric program for silent watch applications. SOURCE: Auburn University.

able research on the materials. The Research Triangle Institute has achieved high figures of merit (ZT ~ 2.4), as reported by *Nature,* and is developing a small thermoelectric system that uses JP-8 as a fuel under the DARPA Palm Power program (Venkatasubramanian et al., 2001). The efficiency of its near-term projected system is 5 percent. As discussed in Appendix C (see Figures C-4 and C-5) and in the section on MEMS microturbines, low-efficiency systems must be very lightweight if they are to be implemented. Although the thermoelectric conversion devices may be light weight (~50 g), the fuel tank, insulation, and combustion components will increase the weight of the system. Therefore such devices may be useful for 24- and 72-hr 20-W missions only if the efficiency of the full system can be increased more than 10 to 15 percent, which will be difficult to achieve. The low TRL of the systems (2) prevents their consideration in the technology assessments of Chapter 2.

Thermophotovoltaics

Thermophotovoltaics, TPV, was considered a potential energy conversion mechanism for the soldier in *Energy-Efficient Technologies* (NRC, 1997). Advances in the art since that report have been steady, but no major breakthroughs have been reported. In a recent paper, R.R. Siergiej et al. (2002) reported the 20 percent conversion of thermal energy incident on cells to electric energy. The main programs in place are classified Navy programs and NASA programs aimed at deep space applications. Both the Navy and NASA programs use heat generated by nuclear sources.

There has been little progress in the development of fueled TPV systems, and, no system has yet been built with an end-to-end efficiency of more than a few percent. TPV is still not a viable candidate for soldier power due to the low efficiency of the technology.

Energy Harvesting

Energy-harvesting technologies were covered in detail in Appendix C of *Energy-Efficient Technologies* (NRC, 1997) and in a recent report from Pacific Northwest National Laboratory (De Steese et al., 2000). While energy-harvesting technologies are compelling because the energy source is inexhaustible, their specific power/power density capabilities must be realistically assessed. In general, energy-harvesting methods have low power and/or energy, making them inappropriate for standard Army soldier needs. The weight of the power conversion device can also make the specific power of the system poor.

For example, a heel strike system capable of generating 2 W (1 W per boot) would yield 16 Wh of energy if the soldier walked for 8 hours. This corresponds to 10 percent of the capacity of the BA-5590 primary battery or an equivalent weight savings of 100 per 8 hours of walking. The heel strike mechanism for extracting mechanical energy from walking might increase fatigue. This and other negative impacts on soldier performance must be taken into account. Obviously, heel strikes would be inappropriate for missions that do not include walking. However, they might find specialized functions—for instance, as the power source for sensors on a soldier's boot that provide information about chemical, nuclear, and other hazards while the soldier is on patrol.

Another option is hand-cranking, several minutes of which can generate 10 to 100 W, enough to power a radio for a limited time. The requirement for hand cranking might have an impact on a stealth mission—or be an inconvenient exercise in the heat of battle or in the fatigue that ensues. Such devices are not lightweight and could be burdensome.

Other proven energy-harvesting methods involve thermoelectric and solar energy. Clearly, a soldier with thermoelectric devices might be able to take advantage of gradients between his/her body temperature and the ambient. Or, photovoltaic devices on the uniform could capture energy from the Sun.

It should be noted that many of the energy harvesting systems discussed above are at high TRLs: Heat/thermoelectric powered watches (from Citizen among others), hand-cranked radios, phones, and flashlights (from Freeplay among others), and solar cells are all commercially available. Even rudimentary piezoelectric heel strike devices have been demonstrated and are likely to further improve (Pelrine et al., 2001). Unfortunately, the weight, cost, and reliability of such devices do not make them viable for applications requiring 20 W or more power. Also, their power and energy can be variable and intermittent and might not be attractive for soldiers facing life-or-death situations.

For all these reasons, energy harvesting methods were not considered in Chapter 2 for technologies that require average power of 20 W or more. However, energy harvesting and human powered systems will become much more attractive in the overall future if the demand for soldier systems can be reduced to 2 W or less, as discussed in Chapter 7.

REFERENCES

Published

Abraham, K.M., and Z. Jiang. 1996. A polymer electrolyte-based rechargeable lithium/oxygen battery. J. Electrochemical Soc. 143: 1.

Acker, W. 2003. Military communications applications for fuel cells. Presented at the NDIA Tri-Services Power Expo, July 15-17, Norfolk, Va.

Amatucci, G., inventor. Telcordia Technologies, Inc., assignee. June 26, 2001. Rechargeable Hybrid Battery/Supercapacitor System. U.S. Patent 6,252,762.

Amatucci, G., inventor. Telcordia Technologies, Inc., assignee. February 11, 2003. High Energy Density Hybrid Battery/Supercapacitor System. U.S. Patent 6,517,972.

Arana, L.R., S.B. Schaevitz, A.J. Franz, M.A. Schmidt, K.F. Jensen. 2003. A microfabricated suspended-tube chemical reactor for thermally efficient fuel processing. J. Microelectromechanical Sys. 12(5): 600-612.

Ball Aerospace. 1998. PPS-50 Operator's Manual. Boulder, Colo.

Ball Aerospace. 2003. Portable Fuel Cells and Power Management. Available at www.ball.com/aerospace/pdf/pps.pdf. Last accessed on January 28, 2004.

Ball Aerospace. 2004. Palm Power DMFC 20-watt system development: DMFC 20-watt system performance and delivery. Presentation by Jeffrey Schmidt to the DARPA Palm Power 2004 Program Review Meeting. Orlando, Fla. February 9-11.

Bar-Cohen, Yosef. 2001. Electroactive polymer actuators and devices. Smart Structures and Materials 2001, Yoseph Bar-Cohen, ed. Pp. 148-156.

Conway, B.E. 1999. Electrochemical Supercapacitors. Kluwer Academic/Plenum Publishers.

Cooper, R., and P. Cox. 2003. PolyFuel—Power for the Wireless World. July 2003.

Crumm, A. 2004. Microtubule solid oxide fuel cells. Presented by Aaron Crumm, Adaptive Materials, to the DARPA Palm Power 2004 Program Review Meeting. Orlando, Fla. February 9-11.

Deluga, G.A., J.R. Salge, L.D. Schmidt, and X.E. Verykios. 2004. Renewable hydrogen from ethanol by autothermal reforming. Science 303(5660): 993-997.

De Steese, John G., Donald J. Hammerstrom, and Lawrence A. Schienbein. 2000. Electric power from ambient energy sources. PNNL Report 13336, September. Available from the National Technical Information Service, U.S. Department of Commerce.

DOE (Department of Energy). 2002. Fuel Cells Handbook. 6th ed. Available online at http://www.seca.doe.gov/refshelf.html. Accessed August 2004.

Ehrich, F.F., and S. A. Jacobson. 2003. Development of high-speed gas bearings for high-power density microdevices. Journal of Engineering for Gas Turbines and Power-Transactions of the ASME 125(1): 141-148.

Epstein, A.H., S.D. Senturia, I.A. Waitz;, J.H. Lang, S.A. Jacobson, F.F. Ehrich, M.A. Schmidt, G.K. Ananthasuresh, M.S. Spearing, K.S. Breuer, and S.F. Nagle. 1999. Massachusetts Institute of Technology, assignee. August 3, 1999. Microturbomachinery. U.S. Patent 5,932,940.

Grubb, William T., inventor. General Electirc Company, assignee. May 26, 1970. Process and Apparatus for Electrochemically Oxidizing Alcohol to Generate Electrical Energy. U.S. Patent 3,514,335.

Hickner, M., B. Pivovar, Y. Kim, J. McGrath, and P. Zelenay. 2002. The effects of sulfonated poly(arylene ether sulfone) copolymer proton exchange membrane post treatment on DMFC performance and properties. Proceedings of the Third International Symposium on Proton Conducting Fuel Cells. Salt Lake City, Utah, October 20-24.

Hu, J.L., Y. Wang, D. VanderWiel, C. Chin, D. Palo, R. Rozmiarek, R. Dagle, J. Cao, J. Holladay, and E. Baker. 2003. Fuel processing for portable power applications. Chem. Eng. J. 93(1): 55-60.

JASON. 2003. Portable Energy for the Dismounted Soldier. Study leader: Nathan S. Lewis, JSR-02-135. McLean, Va.: The MITRE Corporation, June 15.

Johnston, J.P., S. Kang, T. Arima, M. Matsunaga, J. Tsuru, and F.B. Prinz. 2003. Performance of a micro-scale radial-flow compressor impeller made of silicon nitride. Proceedings of the International Gas Turbine Congress (IGTC), Tokyo, pp 1-7.

Linden, D., and T.B. Reddy, eds. 2002. Handbook of Batteries, 3rd ed. New York: McGraw-Hill.

Minh, N.A. 1993. Ceramic fuel cells. Journal of the American Ceramics Society 76: 563-588.

MTI (MTI MicroFuel Cell, Inc.). 2003. MTI MicroFuel Cell's chief technology officer to speak at 2003 fuel cell seminar. Press release. Available online at http://biz.yahoo.com/prnews/031030/nyth174_1.html. Last accessed on January 28, 2004.

NRC (National Research Council). 1997. Energy-Efficient Technologies for the Dismounted Soldier. Washington, D.C.: National Academy Press.

NRC. 2003. Future Hydrogen Production and Use. Letter Report. Washington, D.C.: The National Academies Press.

Park, S.D., J.M. Vohs, and R.J. Gorte, 2000. Direct oxidation of hydrocarbons in a solid-oxide fuel cell. Nature 404: 265-267.

Pelrine, Ron, Roy D. Kornbluh, Joseph Eckerle, Philip Jeuck, Seajin Oh, Qibing Pei, and Scott Stanford. 2001. Dielectric elastomers: Generator-mode fundamentals and applications. Proceedings of SPIE 4329.

Raimondeau S, D. Norton , D.G. Vlachos, and R.I. Masel. 2003. Modeling of high temperature microburners. Part 1. Proc. Combustion Inst. 29: 901-907.

Read, J., K. Mutolo, M. Ervin, W. Behl, J. Solfenstine, A. Driedger, and D. Foster. 2003. Oxygen transport properties of organic electrolytes and performance of lithium/oxygen battery. J. Electrochem. Soc. 150: A1351.

Ren, Xiaoming, and Shimson Gottesfeld, inventors. The Regents of the University of California, assignee. October 2, 2001. Enhanced Methanol Utilization in Direct Methanol Fuel Cell. U.S. Patent 6,296,964.

Schmidt, Jeffrey, and Tim Quakenbush. 2002. Fuel for Portable Power Systems Program. Final report. Prepared for the U.S. Army Research Office by Ball Aerospace and Technologies.

Schreiber, J.G. 2001. Assessment of the Free-Piston Stirling Converter as a Long Life Power Converter for Space. NASA/TM–2001-210604, AIAA–2000-3021.

Siergiej, R.R., B. Wernsman, S.A. Derry, R.G. Mahorter, R.J. Wehrer, S.D. Link, M.N. Palmisiano, R L. Messham, S. Murray, C.S. Murray, F. Newman, J. Hills, and D. Taylor. 2002. 20% Efficient InGaAs/InPAs thermophotovoltaic cells. Proceedings of the Fifth Conference on Thermophotovoltaic Generation of Electricity 653(1): 414-423.

U.S. Army. 1993. Front End Analysis of Soldier Individual Power Systems, USA-BRDC-TR//2541. E. Raskovich, ed. Ft. Belvoir, Va.: Ft. Belvoir Research, Development, and Engineering Center.

Venkatasubramanian, R., E. Siivola, T. Colpitts, and B.C. O'Quinn. 2001. Thin-film thermoelectric devices with high room-temperature figure-of-merit. Nature 413: 597-602.

Whittingham, M.S. 1975. Mechanism of reduction of fluorographite cathode. Journal of the Electrochemical Society 122: 526.

Wood, J.G, and N. Lane. 2003. Advanced 35 W free piston Stirling engine for space applications. Proc. Space Technology and Applications International Forum (STAIF) 2003. Albuquerque, N.Mex.: February 2-5.

Zheng, J.P. 2003.The limitations of energy density of battery/double-layer capacitor asymmetric cells. Journal of the Electrochemical Society 150 (4): A484-A492.

Unpublished

Epstein, A.H., S.A. Jacobson, F. Ehrich, Y. Gong, R. Khanna, J. Lang, H. Li, L. Liu, C. Livermore, H.S. Moon, J. Protz, N. Savoulides, M. Schmidt, C. Spadaccini, Z. Spakovszky, M. Spearing, C.J. Teo, I. Waitz, and D. Ward. Centimeter-diameter gas turbine generators for compact power. Poster presentation to Army Collaborative Technology Alliances Conference. May 2003.

Paur, R. Army research initiatives. Presentation by Richard Paur to the committee on August 20, 2003.